AF601469

Advanced Structured Materials

Volume 32

Series Editors

Andreas Öchsner
Lucas F. M. da Silva
Holm Altenbach

For further volumes:
http://www.springer.com/series/8611

Andreas Öchsner · Lucas F. M. da Silva
Holm Altenbach
Editors

Design and Analysis of Materials and Engineering Structures

Editors
Andreas Öchsner
Department of Applied Mechanics
Faculty of Mechanical Engineering
University of Technology Malaysia—UTM
Johor
Malaysia

Holm Altenbach
Lehrstuhl für Technische Mechanik
Fakultät für Maschinenbau
Universität Magdeburg
Magdeburg
Germany

Lucas F. M. da Silva
Department of Mechanical Engineering
Faculty of Engineering
University of Porto
Porto
Portugal

ISSN 1869-8433 ISSN 1869-8441 (electronic)
ISBN 978-3-642-32294-5 ISBN 978-3-642-32295-2 (eBook)
DOI 10.1007/978-3-642-32295-2
Springer Heidelberg New York Dordrecht London

Library of Congress Control Number: 2012949076

Printed on acid-free paper

Springer is part of Springer Science+Business Media (www.springer.com)

Preface

Design and analysis of materials and structures is nowadays an important discipline which enables a better and more reliable application of engineering components. Furthermore, limits of materials and structure can be accurately determined which may influence the design process and result, for example, in much lighter structures than a few decades ago. Much of these advancements are connected with the increased computer power (hardware) and the development of well-engineered computer software. This directly influences the design and analysis process, for instance, based on numerical simulations (e.g. finite element method) or advanced experimental investigations with modern data acquisition and analysis tools.

The fifth international conference on advanced computational engineering and experimenting, ACE-X 2011, was held in Algarve, Portugal, from 3–6 July, 2011 with a strong focus on computational-based and supported engineering. This conference served as an excellent platform for the engineering community to meet with each other and to exchange the latest ideas. This volume contains 12 revised and extended research articles written by experienced researchers participating in the conference. The book will offer the state-of-the-art of tremendous advances in mechanical and civil engineering, ranging from automotive to dam design, transmission towers up to machine design, and examples taken from the oil industry. Well-known experts present their research on damage and fracture of material and structures, materials modeling, and evaluation up to image processing and visualization for advanced analyses and evaluation.

The organizers and editors wish to thank all the authors for their participation and cooperation which made this volume possible. Finally, we would like to thank the team of Springer-Verlag, especially Dr. Christoph Baumann, for the excellent cooperation during the preparation of this volume.

June 2012

Andreas Öchsner
Lucas F. M. da Silva
Holm Altenbach

Contents

Design of Driveline Test Bench for Noise and Vibration Harshness Improvement of Automotive Chassis Components System

Kee Joo Kim, Si-Tae Won, Kyung Shik Kim, Byung-Ik Choi, Jun-Hyub Park, Jong Han Lim, Jun Kyu Yoon, Sang Shik Kim and Jae-Woong Lee

Abstract The test bench for handling the vibration input and output in a driveline is presented in this contribution. In the experiment, the rear subframe and propeller shafts and axle were composed and mounted with rubber mounts each other as a role of vibration absorbing function. For applying the vibration input instead of the torsional vibration effect of an engine, the shaker moved only the upper and lower side excitation was taken. In particular, the torsional vibration due to fluctuating forced vibration excitation across the joint in between driveline and rear subframe was carefully examined. Accordingly, as the joint response was checked from

K. J. Kim (✉)
Department of Automobile Engineering, Seojeong College University, 1049-56 Whahap-ro, Eunhyeon-myun, Gyeonggi-do, Yangjoo-si 482-777, Korea
e-mail: kjkim@seojeong.ac.kr

S.-T. Won
Department of Product Design and Manufacturing Engineering, Seoul National University of Science and Technology, Seoul 139-743, Korea

K. S. Kim · B.-I. Choi
Nano Mechanics Team, Korea Institute of Machinery and Materials, 1 Jang-dong, Yusung-gu, Taejon 305-343, Korea

J.-H. Park
Department of Mechatronics Engineering, Tongmyong University, Busan 608-711, Korea

J. H. Lim · J. K. Yoon
Department of Mechanical and Automotive Engineering, Gachon University, Gyeonggi-do 461-701, Korea

S. S. Kim
School of Nano and Advanced Materials Engineering, Gyeongsang National University, Gyeongnam 660-701, Korea

J.-W. Lee
Research and Development Team, GS-ONE Company, Gyeonggi-do 413-851, Korea

A. Öchsner et al. (eds.), *Design and Analysis of Materials and Engineering Structures*, Advanced Structured Materials 32, DOI: 10.1007/978-3-642-32295-2_1,

experiments, the FE-simulation (finite element simulation) using FRF (frequency response function) analysis was performed. All test results were signal processed and validated against numerical simulations. In the present study, a new test bench for measuring the vibration signal and simulating the vehicle chassis system is proposed. The modal value and the mode shape of all components were analyzed using the model to identify the important components affecting driveline noise and vibration. It can be concluded that the simplified test bench could be well established and be used for design guide and development of the vehicle chassis components for the improvement of NVH (noise and vibration harshness) problems.

1 Introduction

The major excitation caused to a vehicle system is torque fluctuation by engine excitation. In recent years, the driveline vibration problems are emphasized due to increased engine power and overall increased NVH (noise and vibration harshness) refinement demand from customers. In order to detect driveline vibration issues in an early design stage for the improvement of NVH problems, it is necessary to test and simulate the vibration mode generated by the driveline detached from the vehicle, which is composed of the rear subframe and propeller shafts, axle and rubber mounts used for reducing the primary driveline vibration transmitted to the vehicle body. In a conventional driveline test bench [1, 2], a combustion engine or an electric motor is used for the torque input. However, these equipments are very expensive and the unbalanced force of the engine itself can be the undesirable noise source. In this study, therefore, a shaker was used instead of the engine torsional vibration input because it is possible to make a low cost driveline vibration test bench and to exclude the undesirable noise input source. In addition, a simplified driveline test bench was proposed and evaluated.

The use of computer aided engineering (CAE) techniques, such as the finite element method, as tools for prediction of NVH performance has become almost a standard approach in the automotive industry [1, 2]. The appropriate use of simulation tools can lead to a significant reduction in the number of prototypes required to validate design alternatives and search the root causes of NVH issues detected at prototypes. In order to reach this stage, a thorough correlation study between experiments and simulation must be carried out. In this work, the vibration mode of the driveline was checked from experiments and FE-simulation for FRF (frequency response function) analysis was performed. The test results were compared with the simulation results in order to get more advanced and reliable CAE simulation.

2 Experiments and Simulation Procedure

In order to evaluate the NVH characteristic due to the engine rotation and torque variation, the rear sub-frame, propeller-shafts, rear axle and drive-shaft were composed and mounted with rubbers. Since the propeller-shaft and drive-shaft were rotating parts, the sensor adhesion for measuring the vibration was difficult. For example, a special gap sensor or razor sensing system is needed for the measurement of rotation components. For applying the vibration input instead of the torsional vibration effect of an engine, a shaker was used as shown in Fig. 1. Therefore, the rotational component was excluded and only the upper and lower sides excitation due to the shaker was applied in the present study. In particular, the vibration due to fluctuating forced vibration excitation across the joint between the driveline and the rear sub-frame was carefully examined. A LMS Co. Pimento equipment was used for the vibration measurement and a PCB (printed circuit board) triaxial accelerator was used for detecting the acceleration of the excitation.

For applying the torque fluctuation instead of engine excitation, the shaker was used with increasing frequency from 15 to 145 Hz in one step measurement.

CAE computer simulation was performed by the method of modal frequency response function (FRF Function) using the commercial program Nastran Sol. 111. The simulation model was composed of 110,000 numbers of solid and shell elements. The simulated shaker speed was swept in the same type of experiment input frequency from 15 to 145 Hz using response value at all characteristic points. Simulated results were compared with experimentally measured results. Figure 2 shows the test bench of the driveline of the rear chassis system. All components were composed and mounted by rubbers coupling.

3 Results and Discussion

In order to improve the agreement between the predicted and experimental results, the appropriate dynamic joint stiffness of the rubber components and a prediction method of the each dynamic stiffness, which was reflected at each sweeping frequency, should be well matched. In addition, the initial loads which were applied at each component when assembling each component of the vehicle chassis system should be known. However, the magnitude of the loads was not known and the load values were different in the vehicle chassis system. Figure 3 shows the driveline test bench of the rear chassis system. All components were composed and mounted by rubber coupling.

The natural frequency and the mode shape were investigated through the modal test of a simple sub-frame. In the simple component modal test, the sub-frame was pending from an elastic cord with an acceleration sensor and the excitation was applied by an impact hammer and the frequency response was achieved. The component model of the sub-frame as shown in Fig. 4a was treated as a line path

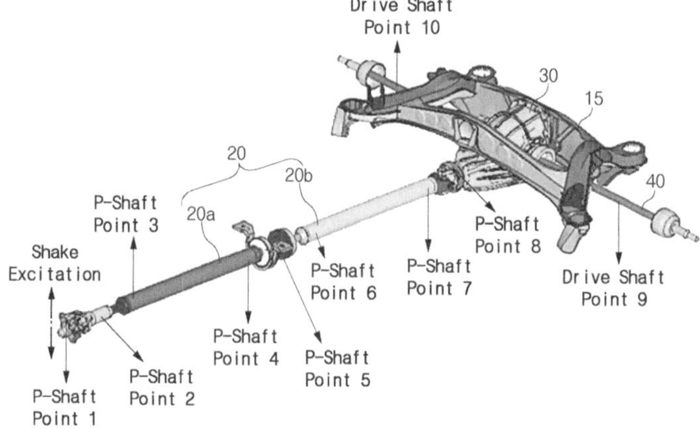

Fig. 1 Schematic model and measuring points of the driveline of the rear chassis system

Fig. 2 Test bench of the driveline of the rear chassis system

and simplified as Fig. 4b in the present study. Figure 5 shows the simplified mode shape and the measured natural frequency of the sub-frame. As shown in Fig. 5, the mode shapes from mode 1 to mode 4 showed a decoupled shape and natural frequency (Hz), respectively.

The finite element model of the sub-frame was made by using two-dimensional shell elements and the natural frequencies were calculated by the Nastran program (Sol. 103 normal mode analysis). Figure 6 shows the mode shape and modal values of the sub-frame from the FE-simulation. From the comparison between Figs. 4 and 5, the mode shapes and the natural frequencies from experiments were in agreement with those from simulation results. Figure 7a, b show the tested and calculated results of vibration animation of the rear chassis system at 27 Hz acceleration. In addition, Fig. 8a, b show the tested and calculated results of

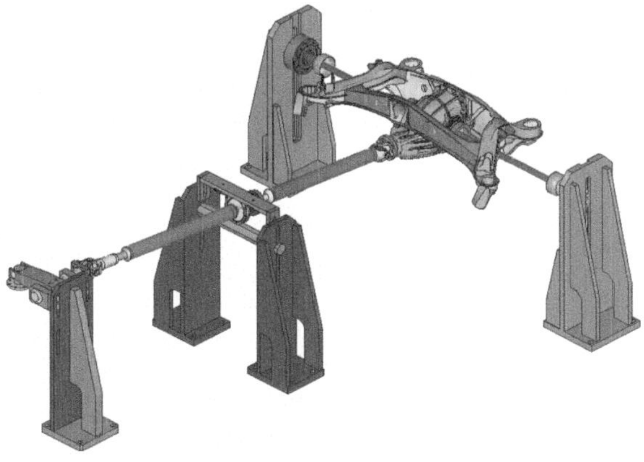

Fig. 3 Model of driveline test bench in rear chassis system

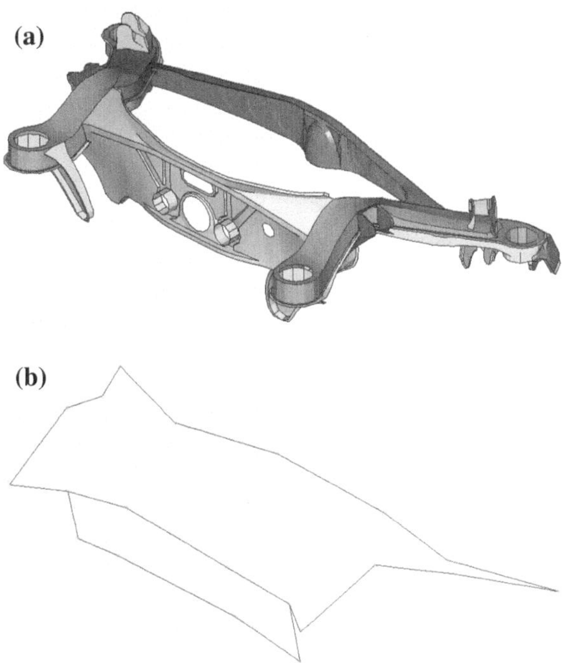

Fig. 4 Example of a simplified rear sub-frame model. **a** Sub-frame model. **b** Simplified sub-frame

vibration animation of the rear chassis system at 33 Hz acceleration. From these results, when compared with experimental results, the calculated results of the system showed reasonably good qualitative agreement.

Figure 9 shows one example of acceleration amplitude (dB) at the shaker point from the experiment. In the case of the shaker, a 130 dB excitation was constantly applied at Z-direction. X-direction had some noise because of equipment limitation.

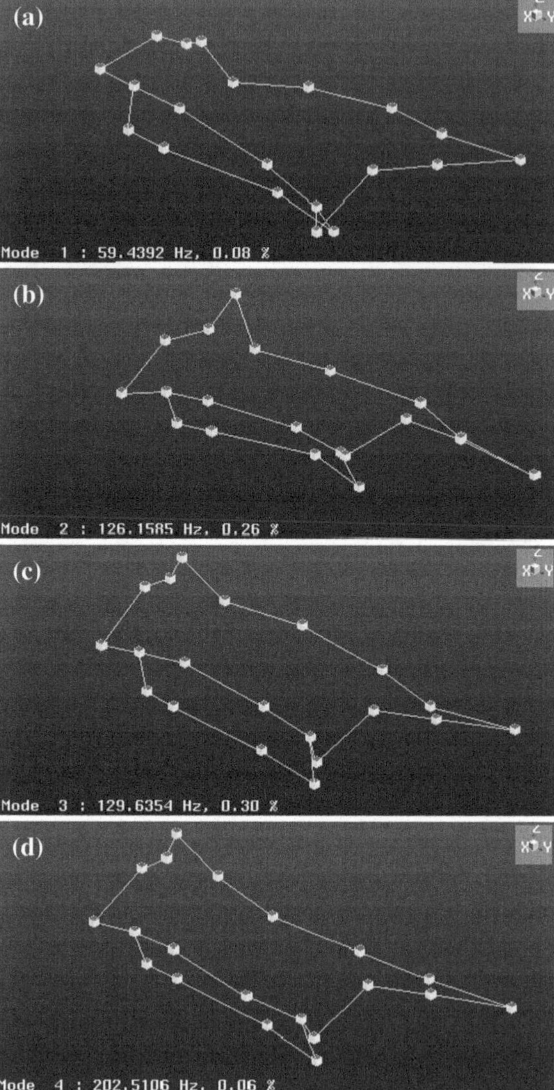

Fig. 5 Modal analysis test results of rear sub-frame. **a** Mode 1 (59.4 Hz). **b** Mode 2 (126.2 Hz). **c** Mode 3 (129.6 Hz). **d** Mode 4 (202.5 Hz)

Figure 10 and 11 show the synthesize representation of measured acceleration at X-direction and Z-direction, respectively, using each points measurement of the propeller shaft based on the shaker excitation. It was confirmed that some special frequencies measured during the experiment had a good agreement with the calculation results, contributed by the resonance. It was also suggested that if the component has higher or lower stiffness without affecting its special resonance frequency, it will have improved NVH characteristic.

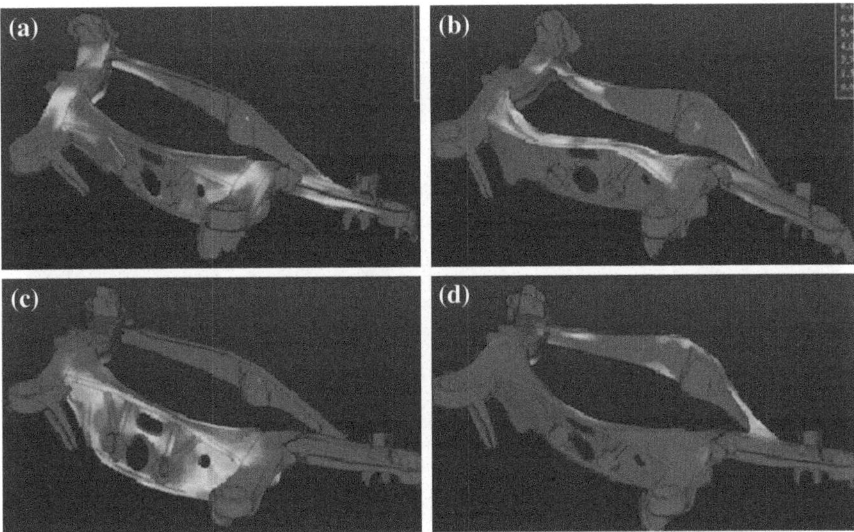

Fig. 6 FE analysis results of rear sub-frame **a** Mode 1 (60.1 Hz). **b** Mode 2 (124.2 Hz). **c** Mode 3 (131.3 Hz). **d** Mode 4 (193.1 Hz)

Each vehicle component has natural frequencies which can be exited by an active source creating the resonance phenomena. This means that the excitation amplitudes are magnified at the natural frequency of the component. Therefore, it is essential to control the interaction of the various resonances by tuning (i.e. frequency shifting) these resonances within the driveline system for better overall performance [3–6]. In order to tune the resonances of the driveline system, it could be considered to change the stiffness of the component and modify the design of the rubber insulators.

In order to improve the prediction capability of the NVH evaluations, however, it is necessary to develop a better test method of the joint parts, considering the rubber properties (static and dynamic stiffness) of each component. Figure 1 shows the measuring points and components name in the rear sub-frame chassis system as previously shown.

Figure 12 shows the ODS (operating deflection shape) of the rear axle roll mode at 66 Hz obtained by the NVH test. It was confirmed from ODS animation that the mode shape at 66 Hz was a roll mode due to the resonance of the axle. Therefore, it was suggested that the stiffness of axle should be higher in order to avoid its special resonance and have an improved NVH performance. Even though the calculated vibration result does not agree well with the measured values quantitatively, the simulation results show that the modal value and mode shape due to the special resonance of the chassis components are considerably similar to the experimental results and could be characterized.

When compared with experimental results, the calculated FRF analysis of the system showed reasonably good qualitative agreement. Further analysis showed

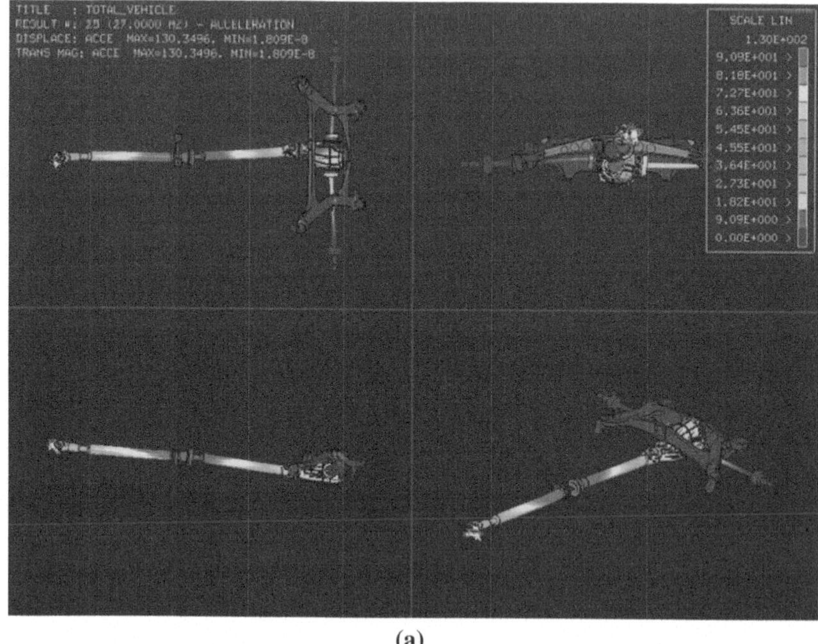

(a)

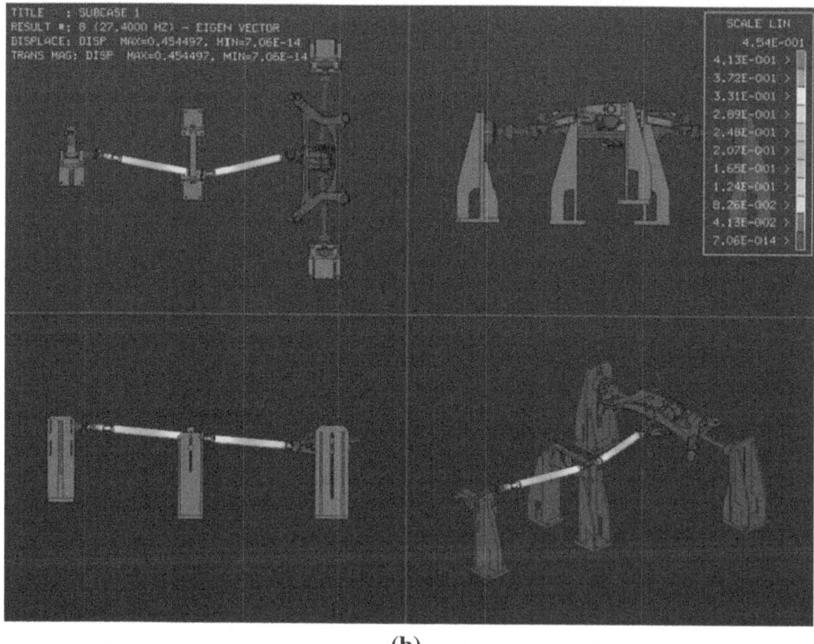

(b)

Fig. 7 Vibration results of the rear chassis system at 27 Hz **a** test and **b** calculation results

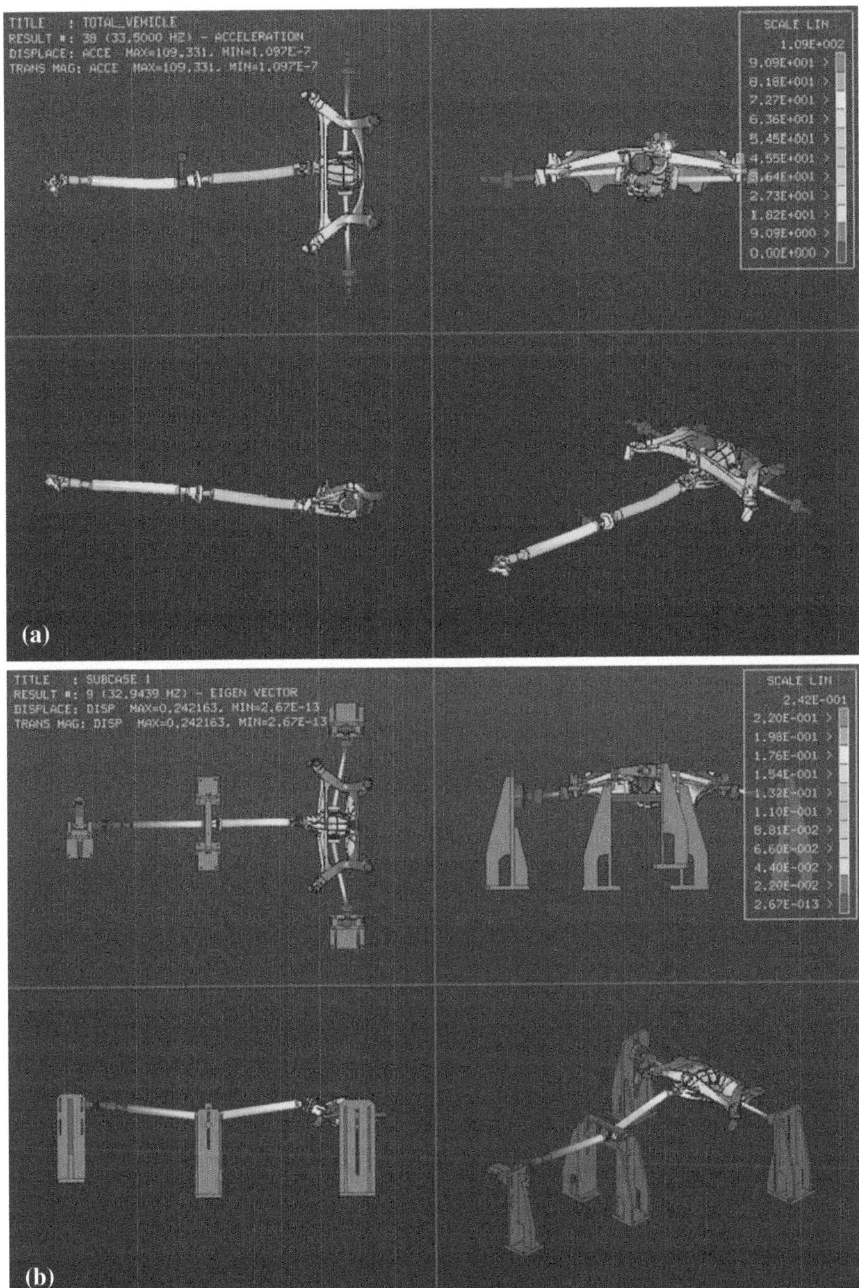

Fig. 8 Vibration results of the rear chassis system at 33 Hz **a** test and **b** calculation results

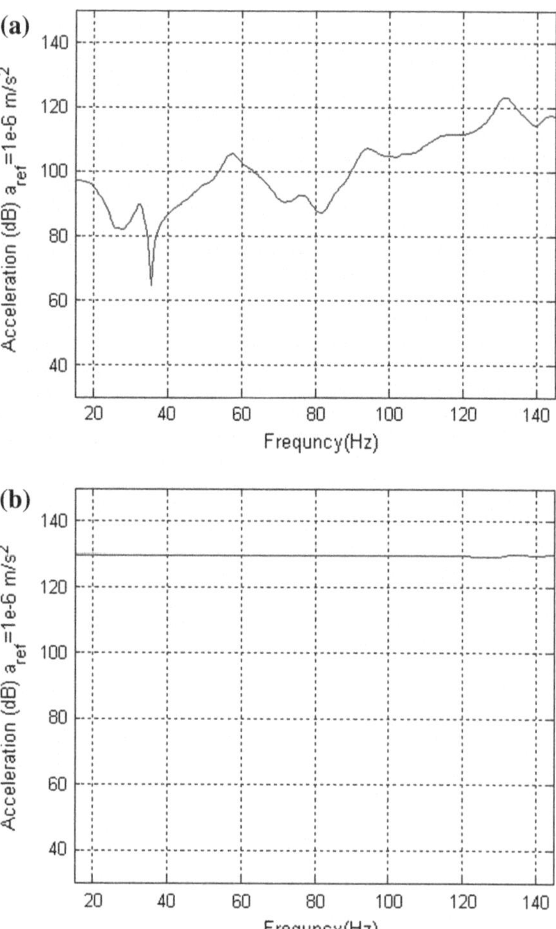

Fig. 9 **a** X-direction and **b** Z-direction acceleration amplitude (dB) at the measuring point of shaker

that the modal value and mode shape simulation of each driveline single components was quantitatively in good agreement with the experimental results. However, the simulated results of the driveline system should be compensated by the joint parts. The joint response by rubber insulators was the main cause of the deviation from the test bench results of the driveline system. The reason of this deviation due to the simulated joint response of rubber components can be explained as follows. One thing was caused by the initial (residual) load which occurred from assembling each rubber components for composing and insulating the chassis system. The other was caused by unknown parameters of the dynamic stiffness in the variable conditions of the experiment which was performed with varying frequency from 15 to 145 Hz at one step. Therefore, it was needed to offset this deviation by a special weight value (e.g. dynamic stiffness ratio). In order to improve the correlation between the predicted and experimental results, a further analysis about the prediction method of each dynamic stiffness of rubber

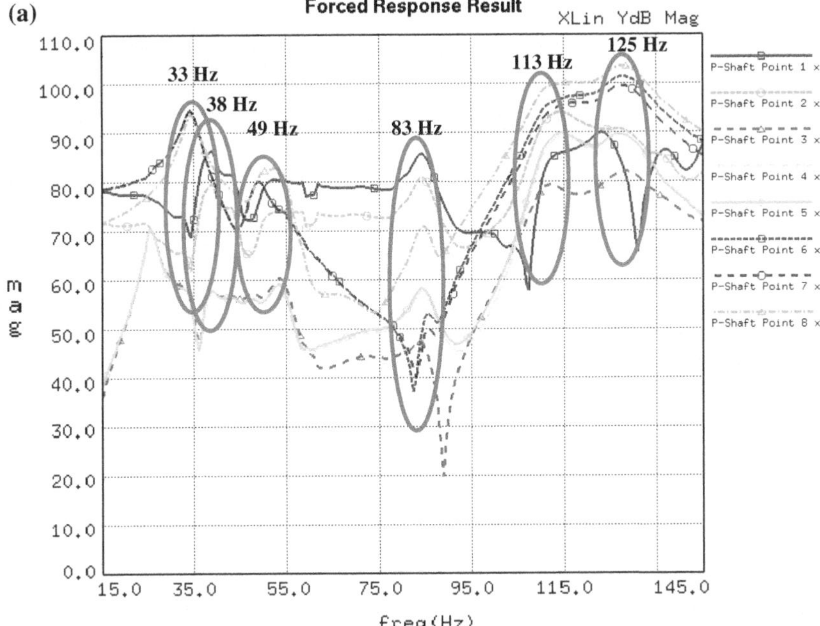

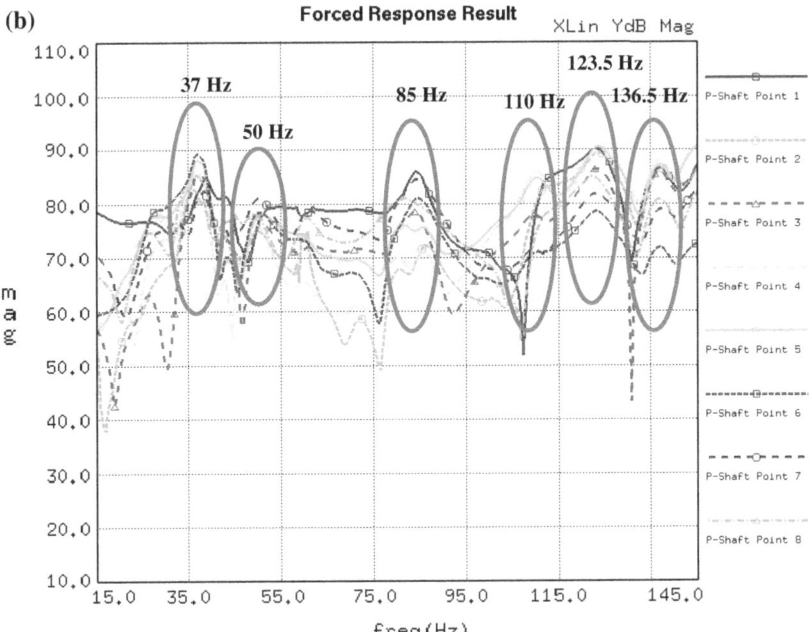

Fig. 10 Rear chassis vibration results (Propeller-shaft) at acceleration (X-direction). **a** FE calculation results. **b** Experimental results

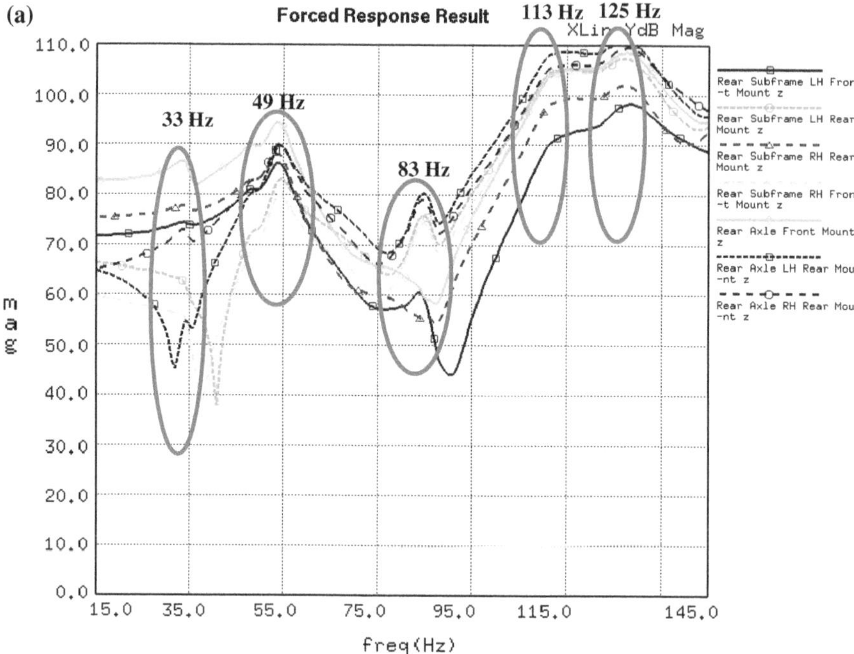

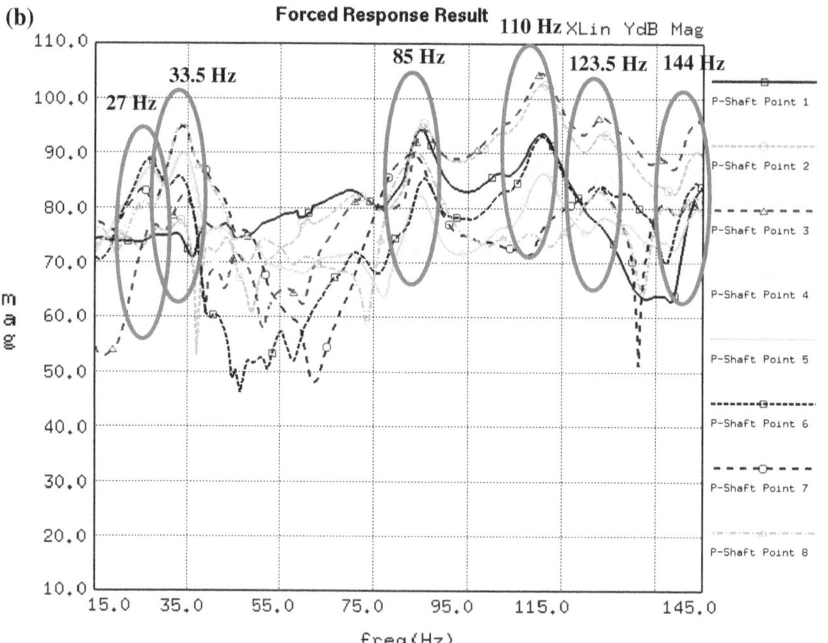

Fig. 11 Rear chassis vibration results (Propeller-shaft) at acceleration (Z-direction). **a** CAE calculation results. **b** Experimental results

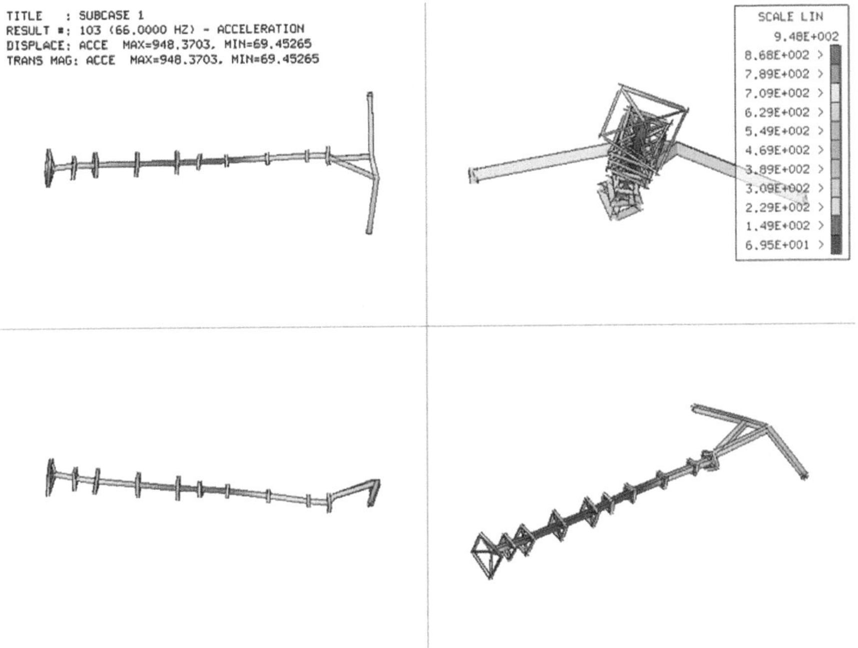

Fig. 12 Operating deflection shape of the rear axle roll mode at 66 Hz

joint parts, which was reflected at each sweeping frequency, should be developed. However, the initial load (remaining pre-load) applied to the joint part when assembling each component of the vehicle is very difficult to be predicted.

4 Conclusion

NVH behavior of the propeller-shaft and drive-shaft by torque fluctuation can be appropriately evaluated from a driveline test bench. When compared with experimental results, the calculated FRF analysis of the system showed reasonably qualitative agreement. However, the calculated results of the driveline system have a minor deviation. This deviation could be supposed by 2 reasons from the joint response of rubber components. One thing was caused by the initial (residual) load which occurred from assembling each component for composing the chassis system. The other was caused by unknown parameters of the dynamic stiffness in the variable conditions of the experiment which was performed with sweeping frequency from 15 to 145 Hz in one step. The NVH characteristics of the simple driveline system, which the body and the other parts were excluded from, were evaluated. This evaluation could contribute to the NVH improvement of the vehicle at the early design stage.

Acknowledgments This research was in part financially supported by a grant from the Center for Advanced Materials Processing (CAMP) of the Twenty-first Century Frontier R&D Program funded by the Ministry of Commerce, Industry and Energy (MOCIE), Republic of Korea. Also, this research was partially supported by the Seoul National University and Science Technology research program.

References

1. Steyer, G, Voight, M., Sun, Z.: Balancing competing design imperatives to overall driveline NVH performance objectives. SAE Technical Paper 2005-01-2308 (2005)
2. Magalhaes, M, Arruda, F., Filho, J.: Driveline structure-borne vibration and noise path analysis of an AWD vehicle using finite elements. SAE Technical Paper 2003-01-3641 (2003)
3. Exner, W.:NVH Phenomena in light truck drivelines. SAE Technical Paper 952641 (1995)
4. Kim, K. J., Rhee, M. H., Choi, B.-I, Kim, C. W., Sung, C. W., Han, C.-P., Kang, K. W.: Development of application technique of aluminum sandwich sheets for automotive hood. Int. J. Precision Eng. Manuf. **10**(4), 71–75 (2009)
5. Kim, K.J., Won, S.T.: Effect of structural variables on automotive body bumper impact beam. Int. J. Auto. Tech. **9**(6), 713–717 (2008)
6. Kim, K. J., Sung, C. W., Baik, Y. N., Lee, Y. H., Bae, D. S., Kim, K.-H., Won, S. T.: Hydroforming simulation of high-strength steel cross-members in an automotive rear subframe. Int. J. Prec. Eng. Manuf. **9**(3), 55–58 (2008)

A New Procedure for the Determination of the Main Technology Parameters of Rolling Mills

Jan Valíček, Jana Müllerová, Veronika Szarková, Krzysztof Rokosz, Czesław Łukianowicz, Dražan Kozak, Pavol Koštial and Marta Harničárová

J. Valíček (✉)
Institute of Physics, Faculty of Mining and Geology, RMTVC,
Faculty of Metallurgy and Materials Engineering, VŠB - Technical University of Ostrava,
17. Listopadu 15/2172, 708 33 Ostrava-Poruba, Czech Republic
e-mail: jan.valicek@vsb.cz

J. Müllerová
Department of Fire Engineering, Faculty of Special Engineering,
University of Žilina, ul. 1. Mája 32, 01026 Žilina, Slovak Republic
e-mail: jana.mullerova@fsi.uniza.sk

V. Szarková
Institute of Economics and Control Systems, Faculty of Mining and Geology,
RMTVC, Faculty of Metallurgy and Materials Engineering,
VŠB - Technical University of Ostrava, 17. Listopadu 15/2172, 708 33 Ostrava-Poruba,
Czech Republic
e-mail: veronika.szarkova@vsb.cz

K. Rokosz · C. Łukianowicz
Division of Electrochemistry and Surface Technology, Koszalin University of Technology,
RACŁAWICKA 15–17, 75-620 Koszalin, Poland
e-mail: rokosz@tu.koszalin.pl

C. Łukianowicz
e-mail: lukianowicz@tu.koszalin.pl

D. Kozak
Faculty of Mechanical Engineering, JJ Strossmayer University of Osijek,
Slavonski Brod, Croatia
e-mail: dkozak@sfsb.hr

P. Koštial
Faculty of Metallurgy and Materials Engineering,
VŠB - Technical University of Ostrava, 17. Listopadu 15/2172,
70833 Ostrava-Poruba, Czech Republic
e-mail: pavel.kostial@vsb.cz

M. Harničárová
Nanotechnology Centre, VŠB - Technical University of Ostrava,
17. Listopadu 15/2172, 708 33 Ostrava-Poruba, Czech Republic
e-mail: marta.harnicarova@vsb.cz

A. Öchsner et al. (eds.), *Design and Analysis of Materials and Engineering Structures*, Advanced Structured Materials 32, DOI: 10.1007/978-3-642-32295-2_2,

Abstract Nowadays, approximately 90–95 % of metals are processed by cold rolling. There has been a substantial increase in demand for utility properties as well as for reducing production costs. These objectives cannot be achieved without a high degree of automation, control and monitoring throughout the manufacturing process. These qualitative changes require rather deep and comprehensive theoretical and metallurgical–technological knowledge of operators in the field of design, research and production of rolled steel sheets, which is needed for further development in rolling steel. A continuous quality control of material and surface during the rolling process is a part of these tasks and is associated with providing the full automation of rolling mills. Starting from theoretical foundations, we have developed a new procedure for the determination of main technology parameters of a rolling mill. The main difference between our proposal and current methods of calculation is as follows. Our proposal is based on the knowledge of deformation properties of materials and continuous processes of stress-deformation state and on the knowledge of reductions in different stages of rolling. Current procedures are on the contrary based on static calculations using the geometry of the system—working roll and instantaneous sheet metal thickness in a gap between the rollers. In doing so, the calculations almost ignore the real stress—deformation properties of rolled metal sheets, optimal transmission rate of deformation in the material at the given speeds of rollers and the given main rolling force. We are concerned with the optimum balanced system: main rolling force—rolling speeds, or transmission rate of deformation in the material. This procedure allows us to achieve a significant increase in operational performance as well as in rolling process quality.

1 Introduction

Rolling means a continuous process in the course of which the material being formed deforms between the working rolls under conditions of prevailing all-around pressure. The material being rolled deforms between the rolls, the height decreases, the material elongates and simultaneously widens, and also the speed at which the material being rolled leaves the rolling mill changes [5].

Continuous cold rolling is an important process in the metallurgical industry. Its performance influences directly the quality of a final product. This process is a very complicated system, including many multidisciplines, such as: computing technique, automatic control, mechanics, material engineering, and others. Any correctly designed recovery of the system may bring a large financial gain, consisting in an increase in performance, quality and overall business competitiveness, to a company [1, 2, 3, 4].

The goal of metal cold rolling process is to manufacture very high quality flat rolled products from metals, where the high quality means that the final product has the required geometry, including sheet flatness [6, 7, 8]. The issue of longitudinal cold rolling and its influence on the topography of surface of rolled sheets

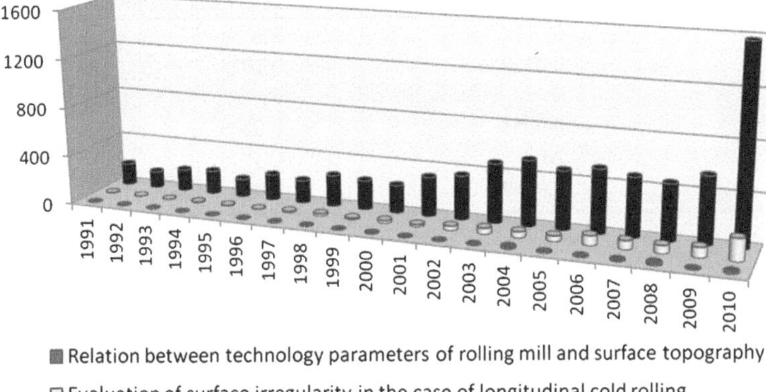

Fig. 1 Overview of world literature in the SCIRUS database

is graphically represented in Fig. 1. It shows this issue dealt with in the world literature in the course of last 20 years with a view to presenting the numbers of publications in individual study areas. Research into the area of relations between technology parameters of the rolling mill and the topography is done very rarely owing to comprehensive complexity of solving.

2 Experimental Procedure

Low carbon steel of type PN EN 10263-2:2004 with the Young's modulus $E_{mat} = 126$ GPa and the yield strength of 310 MPa was chosen as an initial material for the realization of experiments. The chemical composition is given in Table 1. Deep drawing quality steel of U.S.Steel Košice, Ltd. production of type KOHAL 697 with the Young's modulus $E_{mat} = 186.7$ GPa and the yield strength of 390 MPa was chosen as the second material being used in experiments. Its chemical composition is given in Table 2.

Chemical analysis was performed by a glow discharge optical emission spectrometer LECO GDS 750.

The material investigation consisted of the metallographic analysis. The metallographic analysis of the samples were done perpendicularly to the longitudinal axis of the samples. For a visual observation of the samples there was used the inverted microscope for transmitted light GX51 with the maximum magnification of 1000x. Figure 2 shows an example of the microstructure of low carbon steel of type PN EN 10263-2:2004.

The analysis of micro purity does not show any significant contamination by non-metallic additives. There were observed globular-oxide-type inclusions. The microstructure of deep-drawing steel consists of ferrite, pearlite and cementite

Table 1 Chemical composition of the used low carbon steel PN EN 10263-2:2004

Fe	C	Mn	Si
99.620	0.0228	0.1928	0.0117
P	S	Cr	V
0.0081	0.0053	0.0209	0.0018
Cu	Al	Co	As
0.0275	0.0562	0.0075	0.0047

Table 2 Chemical composition of the used deep-drawing steel of type KOHAL 697

Fe	C	Mn	Si
99.5327	0.037	0.251	0.007
P	S	Al	N_2
0.007	0.007	0.051	0.0031
Cu	Ni	Cr	As
0.021	0.009	0.014	0.001
Ti	V	Nb	Mo
0.001	0.001	0.002	0.002

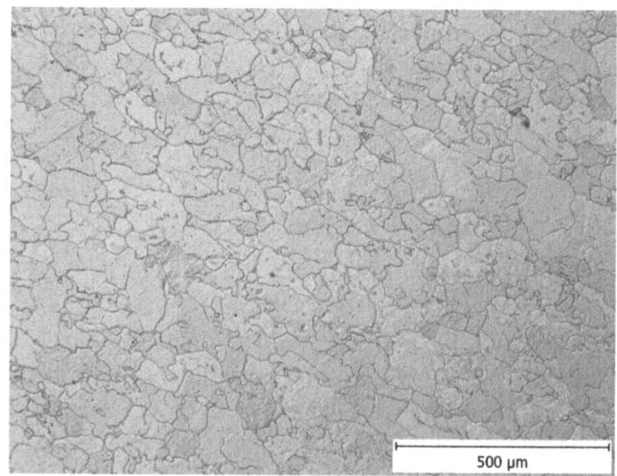

Fig. 2 Microstructure of low carbon steel

(Fig. 3). A ferritic grain is polyedric, equally distributed in cross-sectional area of the investigated material.

In contrast to the existing methods, the final condition of rolled sheet surface topography was taken as an initial input parameter for solving. It is above all the final roughness of sheet surface in relation to the main technology parameters, namely the rolling force F_{roll}, rolling speed v_{roll}, reduction Δh and in relation to sheet materials. Low carbon steel sheets and deep-drawing steel sheets produced by longitudinal cold rolling were measured using an optical profilometer Micro-Prof FRT. Samples were placed on a scanning table and an area of about 5×5 mm^2 size (Fig. 4) was scanned by a fixed sensor at a step of 3 µm and a frequency of 1 kHz. In this way, data on the studied surface topography were

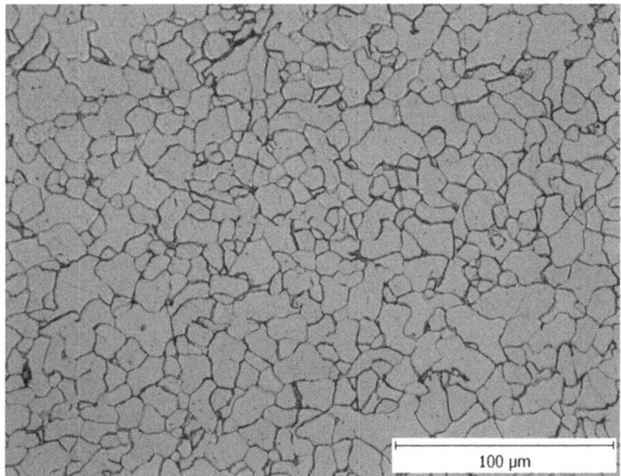

Fig. 3 Microstructure of deep-drawing steel

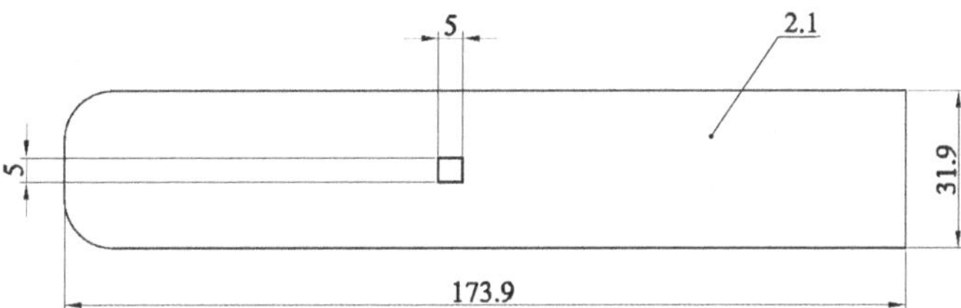

Fig. 4 Diagram of sample 1C with 0.4 mm reduction and delimited 5 × 5 mm^2 measured area

acquired. By the optical profilometer, the 3D topography of surface of measured areas was obtained.

Information on surface irregularities produced by longitudinal cold rolling for individual reductions Δh, i.e. the mean arithmetic deviation of surface profile *Ra*, root mean square deviation of profile *Rq* and the maximum height of profile irregularity *Rz*, is given in Table 3.

On the basis of data acquired from low carbon steel and deep-drawing steel, a graph in Fig. 5 was constructed. It shows that the quality of surface of low carbon steel sheets is higher (lower values of mean arithmetic deviation of surface profile *Ra*) than that of surface of deep-drawing steel sheets.

Surface roughness is expressed implicitly as follows: $Ra = \mathrm{f}\,(F_{roll}, Q_{Froll}, v_{roll}, n_{roll})$. It is necessary to look for relations between technology, material and resultant surface quality.

Using the program Gwyddion, measured areas of individual samples were analysed. Gwyddion is a modular program for analyzing scanning probe microscopy (SPM) data files. Gwyddion is free and open source software, covered by GNU General Public License. In the measured areas of 5 × 5 mm^2, ten equidistant measuring lines at a step 0.5 mm were there. The measuring lines were perpendicular to the direction of rolling. From each measuring line, a signal

Table 3 Relevant standardized parameters of surface texture evaluation for individual low carbon steel and deep-drawing steel samples

Samples from	Sheet designation	Δh (mm)	Ra (μm)	Rq (μm)	Rz (μm)
Low carbon steel	0A	–	0.71	0.97	7.42
	1A	0.4	0.37	0.52	3.39
	2A	0.5	0.43	0.56	3.77
	3A	0.6	0.21	0.28	2.05
	4A	0.8	0.12	0.17	1.60
Deep-drawing steel	0C	0	1.07	1.38	6.94
	1C	0.38	0.76	0.97	4.66
	2C	0.96	0.73	0.94	5.12
	3C	1.27	0.43	0.55	3.02
	4C	1.56	0.61	0.81	4.74
	5C	1.73	0.48	0.61	3.16

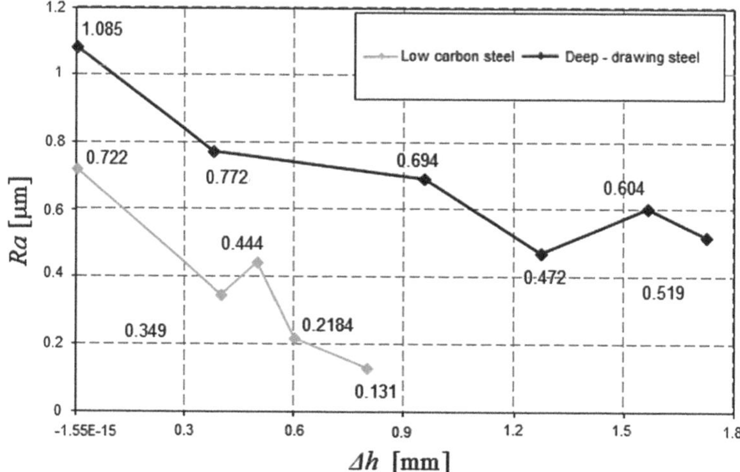

Fig. 5 Graphical representation of dependence of mean arithmetic deviation of surface profile on absolute reduction

carrying the information on the distribution of surface height fluctuations (*RMS*) was obtained (Fig. 6).

The rolling force of a rolling mill DUO 210Sva (testing stand) was already in the case of deep-drawing steel measured by metallic resistance tensometers (see Fig. 7) mounted on this mill in the course of rolling our sheets.

The rolling force was measured at a rolling speed of 0.7 m s^{-1} (see Fig. 8a) in the case of all deep-drawing steel sheets; the rolling force ranged from 67.9 to 145.1 kN. The greater is the used reduction, the higher is the rolling force and also the smoother is the surface of the given rolled product, as graphically represented in Fig. 8b.

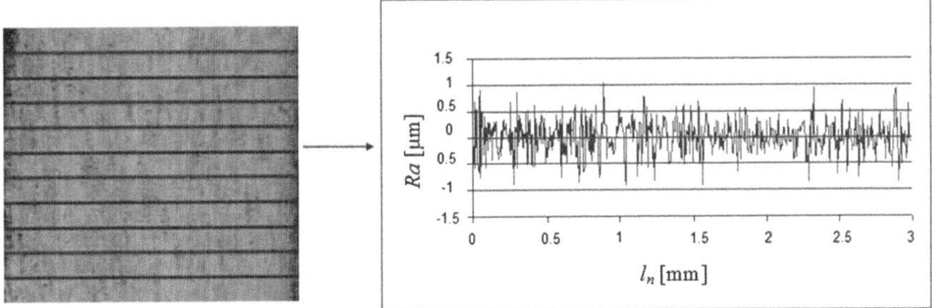

Fig. 6 Measured areas of sample with 0.6 mm reduction in ten equidistant lines (0.5 mm distance) from which signals representing surface irregularities were obtained

Fig. 7 Detail showing the location of tensometers on the rolling mill

In Fig. 8b a difference between the curve representing the dependence of rolling force on reduction for low carbon steel and the curve for deep-drawing steel is clear. This difference is given by different numbers of reductions and reduction values that were not identical for the low carbon and deep-drawing steels (see Table 4).

3 Analytical Processing

For a purpose of analytical processing and generalization, the authors have chosen the concept of regression and correlation analysis. In order to determine the main functions of rolling process there was measured a number of metallic materials with different mechanical parameters. As a reference parameter was chosen the modulus of elasticity and the material constant of plasticity K_{plmat} (1) used in previous works of the authors.

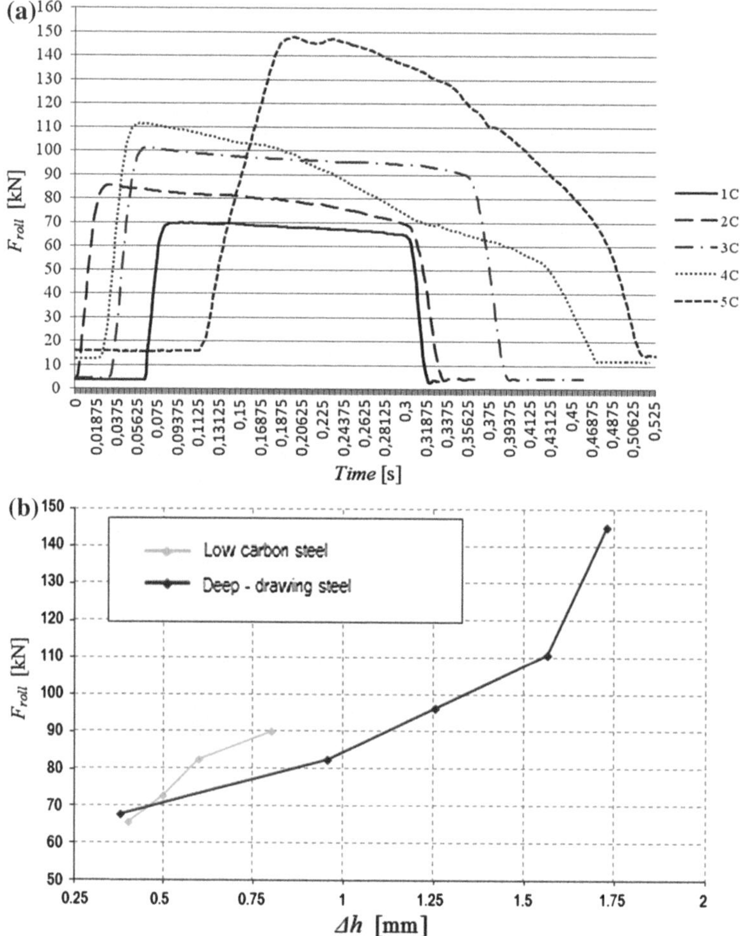

Fig. 8 **a** Graphical representation of rolling force variation in time at 0.7 m s^{-1} rolling speed. **b** Relation between the reduction Δh and the rolling force F_{roll}

$$K_{plmat} = \frac{10^{12}}{E_{mat}^2}\left[\text{MPa}^{-2}\right] \tag{1}$$

It was found that using the index ratio (2) that is incorporated into the regression relations for reduction in thickness, it is possible to distinguish different materials in the process of rolling.

$$I_{kpl} = \sqrt{\frac{K_{plmat0}}{K_{plmat}}}[-] \tag{2}$$

Table 4 Dependence of mean arithmetic deviation of surface profile on rolling force

Samples from	Sheet designation	Δh (mm)	Ra (μm)	F_{roll} (kN)
Low carbon steel	0A	–	0.71	0
	1A	0.4	0.37	65.55479
	2A	0.5	0.43	72.90912
	3A	0.6	0.21	82.71223
	4A	0.8	0.12	90.06656
Deep-drawing steel	0C	0	1.07	0
	1C	0.38	0.76	67.8861
	2C	0.96	0.73	82.5948
	3C	1.27	0.43	96.3955
	4C	1.56	0.61	110.9628
	5C	1.73	0.48	145.1399

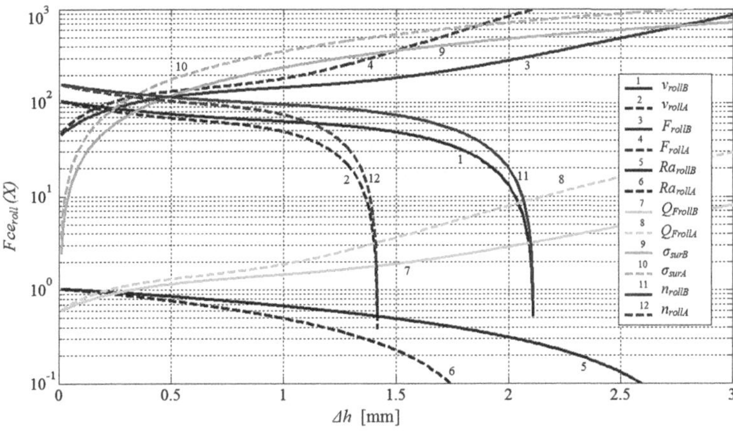

Fig. 9 Mathematical model illustrating the course of the main rolling parameters for the used materials deep-drawing steel (material A, $Ementry = 186.7$ GPa) and low carbon steel (material B, $Ementry = 126$ GPa)

Based on this finding, an algorithm for mathematical modeling in MATLAB has been developed. An example of the course of basic parameters for rolling of the material marked as “A” (low carbon steel) and material “B” (deep-drawing steel) is shown in Fig. 9.

A set of equations for the algorithm (3) and Fig. 9 contains the following variables for material “A” and “B”:

σ_{sA}, σ_{sB}	Surface tension (MPa),
F_{rollA}, F_{rollB}	Rolling force (kN),
n_{rollA}, n_{rollB}	Rotations of the rolls (rot $\min^{-1}$),
v_{rollA}, v_{rollB}	Rolling speed (m $\min^{-1}$),
Q_{FrollA}, Q_{FrollB}	Forming factor (–),
Ra_A, Ra_B	Measured and predicted surface roughness (μm),
Δh	Reduction in thickness (mm),

D_{gr}	Grain size (mm),
S_h	Horizontal projection of the contact surface between the rolled metal and the roll (mm^2),
l_d	Length of deformation (mm),
h_s	Average value of reduction in thickness (mm),
Δh_{maxA}, Δh_{maxB}	Maximum reduction in thickness for material A and B (mm),
D_{roll}	Roll diameter,
b_s	Average width of rolled sheet [8, 6].

The input equations to the algorithm and their exact record (3) are given below:

```
Droll = 0.210; delh = 0.01:0.01:8;
EmentryA = 186700; Kplmato = 28.68873;
EmentryB = 125580; Kplmat = 10.^12/1./Ementry.^2;
Ikpl = (Kplmato/ Kplmat) ^0.5
Vroll = 148.0662 – 132.41266.*(delh* Ikpl) + 160.17705.*(delh* Ikpl).^2 –
        83.55556.*(delh* Ikpl).^3;
Froll = 43.63183 + 329.16274.*(delh* Ikpl) – 413.33298.*(delh* Ikpl).^2 +
        222.8675.*(delh* Ikpl).^3;
Ra = 1.47963 – 0.534 .* (delh* Ikpl);
QFroll = 0.57873 + 2.50374.*(delh*Ikpl) – 3.00062.*(delh*Ikpl).^2 +
 1.78554.*(delh* Ikpl).^3;
SIGsur = 4.66667E – 5 + 354.26751 .* (delh* Ikpl);
nroll = vroll /1./(3.14.*Droll);
Dgrc = 0.05*(0.34843 + 3.30751 .* Rarollm);
Sh = 1000*Froll/1./( Emat^0.5.*QFroll);
ld = Sh/1./bs;
hs = ld/1./QFroll;
MATactual = 1001.
```

The curves of these parameters D_{gr}, S_h, L_d and h_s are not plotted in Fig. 9 to make a clear interpretation.

Figure 9 illustrates the course of rolling functions, showing a high degree of comformity with the measured values given in Tables 1 and 4 and in Figs. 5 and 8 according to individual reductions in thickness of both materials. Using the algorithm (3) that contains the regression equations, it is possible to get the high degree of conformity between the measured and predicted values. The closeness of the resulting values does not exceed 10 %. The model calculation requires to control the rolling speed v_{roll} and to control the rotations of the rolls n_{roll} depending on the increase in these parameters: thickness reduction Δh_i, rolling force F_{roll}, surface strengthening σ_{sur} and forming factor Q_{Froll} while the surface roughness values are decreasing. The surface tension curve σ_{sur} can be considered as a function of strengthening.

4 Conclusion

The theoretical treatment and the basic project concerning the operational application of optical check of mechanical parameters, especially the yield strength of a material of cold rolled sheets, rest conceptually on basic documents and specific operating conditions at a workplace of a specific company. Values of trace roughness *Ra* are input data for worked-out algorithms. Interpretation procedures and their results are conceived to provide not only the check values, i.e. in principle values obtained by defectoscopy, but also to model-predict the instantaneous mechanical condition of the material, capacity-deformation data, and others in the whole rolling process so that they can be used in designer's work both in the stage of project designing the rolling parameters and in the stages of check and control of sheet cold rolling process. Thus the algorithm, which after putting input material data will automatically provide a comprehensive mathematical model of the process in numeral as well as graphic form, is available for theory and practice. The main goal of the presented solutions was to derive the equations for the algorithm (3) according to experimental works. Looking further ahead, the authors will make some adjustments of the algorithm for industrial multistage cold rolling mills, what shall be followed by verification in real operating conditions. Next step is to propose a way to use the algorithm for on-line control and automatization.

Acknowledgments The contribution was supported by the projects RMTVC No. CZ.1.05/2.1.00/01.0040. Thanks also belong to the Moravian-Silesian Region project RRC/04/2010/36 for financial support and IT4 Innovations Centre of Excellence project, reg. no. CZ.1.05/1.1.00/02.0070.

References

1. Herman, J.C.: Impact of new rolling and cooling technologies on thermomechanically processed steels. Ironmak. Steelmak. 28(2), 159–163 (2001)
2. Kenmochi, K., Yarita, I., Abe, H., Fukuhara, A., Komatu, T., Kaito, H.: Effect of micro-defects on the surface brightness of cold-rolled stainless-steel strip. J. Mater. Process. Technol. **69**, 106–111 (1997)
3. Mišičko, R., Kvačkaj, T., Vlado, M., Gulová, L., Lupták, M., Bidulská, J.: Defects simulation of rolling strip. Mater. Eng. 16(3), 7–12 (2009)
4. Valiev, R.Z., Estrin, Y., Horita, Z.: Producing bulk ultrafine grained materials by severe plastic deformation. J. Minerals Metals Mater. Soc. 58(4), 33–39 (2006)
5. Zrník, J., Dobatkin, S.V., Mamuzič, I.: Processing of metals by severe plastic deformation (SPD)—structure and mechanical properties respond. Metalurgija **47**(3), 211–216 (2008)
6. Ginzburg, V.B.: Flat-Rolled Steel Processes. CRC Press, Boca Raton (2009)
7. Pittner, J., Simaan, M.A.: Tandem Cold Metal Rolling Mill Control. Springer, London (2011)
8. Lenard, J.G.: Primer on Flat Rolling. Elsevier, London (2007)

Observation of Damage to Materials for Educational Purposes at the BSc Level

Ivica Kuzmanić and Igor Vujović

Abstract This chapter presents the results of education in the science and technology of materials at the Maritime Faculty at the BSc level. The goal is the experimental observation of mechanical damage to materials and the impact of such damage on the electrical properties of materials. Since small faculties cannot afford to buy devices used in nanotechnology, the chapter introduces optical microscopy as a cheaper way of introducing micro and nano world to students. Experiments presented to students include the influence of corrosion and cracks on the properties of materials. It is also shown that students preferred this way of education to the classical lessons and seminar works. The examples of damage to materials obtained by optical microscope are linked to the theory as an example of the learning process.

1 Introduction

Damage to and degradation of materials play a vital role in the occurrence of problems in the maintenance of industrial, traffic or any other technical systems. Therefore, it is important to integrate the awareness of the problem in the educational curriculum. Although we are primarily interested in materials used in electrical

I. Kuzmanić (✉) · I. Vujović
University of Split, Zrinsko-Frankopanska 38, 21000, Split, Croatia
e-mail: ikuzman@pfst.hr

I. Vujović
e-mail: ivujovic@pfst.hr

A. Öchsner et al. (eds.), *Design and Analysis of Materials and Engineering Structures*, Advanced Structured Materials 32, DOI: 10.1007/978-3-642-32295-2_3,

engineering, the chapter is also applicable to other technical areas, i.e. mechanical engineering.

Since it is impossible for small educational environments to implement expensive equipment [1, 2] in the educational processes at the BSc level, lecturers have to improvise to maintain the level of knowledge which does justice to the name of the institution and to prepare students for future carriers.

Materials used in electrical engineering can be divided into electrical (engineering) materials, construction materials and additional materials [3]. All are influenced by the passage of time and subject to damage. Degradation of materials can be caused by a variety of reasons, such as thermal [4], mechanical [5], etc. Damage can result from physical or chemical causes and occur under the influences of the micro or macro world. If a material is damaged or degraded by environmental causes, defects (such as dislocations, cracks, cavities and lack of homogeneity) will occur [6, 7]. However, defects may occur in materials either during manufacture or normal use. Fracture mechanics plays a very important role in the analysis of the safety of the final product. Degradation of materials due to both normal aging processes and operating stresses causes fractures and mechanical failures. The likelihood of occurrence of critical stress around defects in the production process is higher than at any other position [8].

2 Theory Linked with Experiment for Students

In order to link the theory and laboratory exercises, the problem must be introduced in lectures. To "measure" the influence of mechanical damage on electrical properties, a stress (safety) intensity factor is introduced.

A stress intensity factor is a measure of mechanical reliability of products. The obtained results suggest that the existence of a crack in a stressed piezoelectric material leads to a stress redistribution and concentration, and the tensors of mechanical and piezoelectric stress at the crack tip become infinitely large. This "point of stress singularity" occurs at the point most sensitive to fracture [8]. Electrical and electronic products can be relied on only if there is no danger of failure. Therefore, it is important to link mechanical damage and electrical properties of materials.

Types of mechanical deformation are shown in Fig. 1. Modes are defined by coordinates r, φ and z in a polar-cylindrical coordinate system and shifting of the parts of the material in axes directions by u, v and w.

Mode I is defined by the separation of the fracture surface symmetrically with regard to the primary crack plane (so-called crack opening): $u = u(r, \varphi), v = v(r, \varphi)$, $w = 0$. Mode II is defined by the movement of one crack surface in relation to the other in the same plane, but in opposite directions (shear): $u = u(r, \varphi), v = v(r, \varphi)$, $w = 0$. Mode III is defined by the movement of one fracture surface along the front of the crack (shear out of plane): $u = 0$, $v = 0$, $w = w(r, \varphi)$.

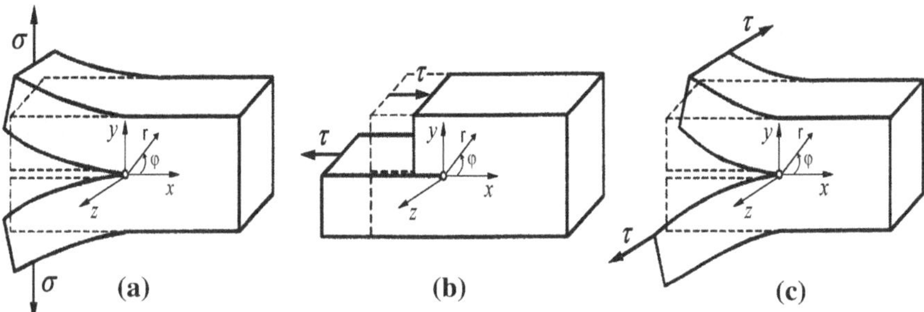

Fig. 1 Crack deformation modes: **a** I, **b** II, **c** III

Every arbitrary deformation can be depicted by the combination of theses modes, i.e. when the crack front is a spatial curve, in an arbitrary point, all three modes can be present. In a polar-cylindrical coordinate system (r, φ, z), general expressions for components of the tensor of mechanical stress state are expressed as [9, 10]:

$$\sigma_{i,j}(r,\varphi,z) = \frac{1}{\sqrt{2 \cdot r \cdot \pi}} \cdot \left[K_I(z)\sigma^{I}_{i,j}(\varphi) + K_{II}(z)\sigma^{II}_{i,j}(\varphi) + K_{III}(z)\sigma^{III}_{i,j}(\varphi)\right] \tag{1}$$

provided that K_{I}, K_{II}, K_{III} are stress intensity factors for modes I, II and III, σ corresponding stress and r the distance from the marked point (see Fig. 1). In the vicinity of the crack, Hooke's law is inapplicable, because of the plastic area in which nonlinear effects are dominant [11]. However, if the characteristic zone dimensions are very small compared to the size of the initial crack, the linear elastic fracture mechanics (LEFM) in an analysis of stress and strain state of the crack tip is possible. In that case, the stress intensity factor is a measure of stress magnitude in the vicinity of the tip of the crack. It is defined for known stress state and geometry.

In order to link the mechanical stress to electrical properties [6, 12, 13], it is necessary to have a measurable electrical value. In the experimental case, surface and volume electrical resistance are such measures. It can be intuitively presumed that the electrical resistance of the material would change if the material is degraded. However, degradation of the insulator leads to smaller electrical resistance. Vice versa, degradation of the conductor material means an increase in electrical resistance. For example, corrosion of copper wires increases electrical resistance, because copper-oxide is a semiconductor.

To derive the formula for electrical resistance in an insulator, students must realize that conduction current exists after the process of polarization. The polarization occurs in the dielectric material in a constant electric field. The process of polarization takes only a nanosecond to be over. The electrical resistance of the material, in which conduction current flows, can be expressed as [14]:

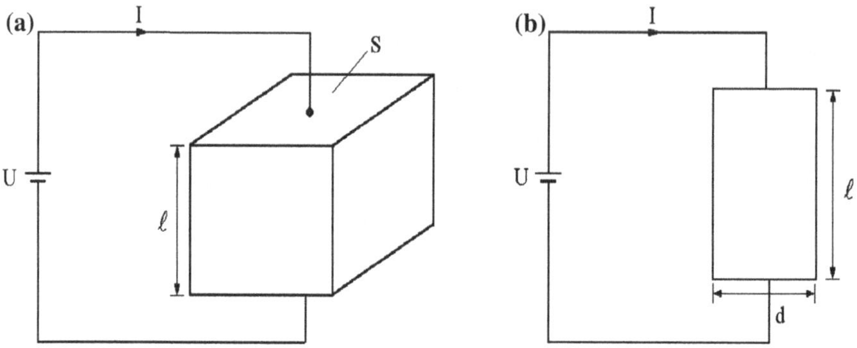

Fig. 2 Illustration of the volume **a** and surface **b** resistance measurement [14]

$$R = \frac{U}{I_c} \tag{2}$$

where U is the constant voltage loaded to the material sample, I_c is the conduction current and R is the electrical resistance. The conduction current, I_c, consists of two components when talking about solids. These components are surface, I_{cs}, and volume, I_{cv}:

$$I_c = I_{cs} + I_{cv} \tag{3}$$

Dividing results in Eq. (3) by U gives:

$$\frac{I_c}{U} = \frac{I_{cs}}{U} + \frac{I_{cv}}{U} \tag{4}$$

It is possible to measure both the volume and the surface resistance (see scheme in the Fig. 2).

If we define volume and surface resistance as:

$$R_v = \frac{U}{I_{cv}} \tag{5}$$

$$R_S = \frac{U}{I_{cs}} \tag{6}$$

and combining Eqs. (4), (5) and (6), the expression for the material's resistance is obtained:

$$\frac{1}{R} = \frac{1}{R_v} + \frac{1}{R_S} \tag{7}$$

Therefore, the insulator (dielectric) material can be replaced by two resistors in parallel, $R_v \| R_s$ and can be expressed as:

$$R_v = \rho_v \cdot \frac{\ell}{S} \tag{8}$$

$$R_S = \rho_S \frac{\ell}{S} \tag{9}$$

The volume electrical resistance depends on: temperature, humidity, additives, strength of the electric field, etc. The surface electrical resistance depends on: environmental humidity, surface purity, polarization, etc.

3 Experiment and Results

Results presented in this section have been obtained by students who have chosen this type of examination. In the initial survey, they represented approximately 40 % of the class at the beginning of the academic year. Other students in the survey have chosen different examination: final exam 20 %, tests and quizzes 20 %, seminar works 20 %. However, about 20 % more chose experiments instead of the final exam during the academic year and similarly 20 % changed from tests to seminar. Finally, 60 % of the students used experiment for examination as an alternative to seminar work, while 40 % opted for classic seminar work at the end of the semester.

Experiment includes:

- detection of suspicious parts of a material,
- observation by optical microscopy, and
- conclusion about mechanical damage and its influence on electrical properties.

Detection of suspicious parts of a material is performed by a simple technique of observation performed in practice at the regular bases. The suspicious material is put on the microscope pad and observed under different amplifications.

Paper damage reduced the surface resistance of the paper, measured by an insulator tester. Paper cut in electrical insulation can cause an insulation tear.

If the CD surface was damaged unrecoverably, the computer would not have read it. Therefore the damage of the CD surface was too small to influence the computer performance. However, CD surface damage is a very large problem. Namely, even during normal operation, the laser beam unintentionally damages the surface. Thus the surface will become unreadable after a number of readings. Furthermore, the CD surface is sensitive to accidental small cuts or touches with edges of even a CD box.

In this chapter, several materials are presented:

- paper from electrical insulations (electrical (engineering) application, electrical machines),
- metal (conductor), which is corroded and
- CD surface (optical memory, computer industry).

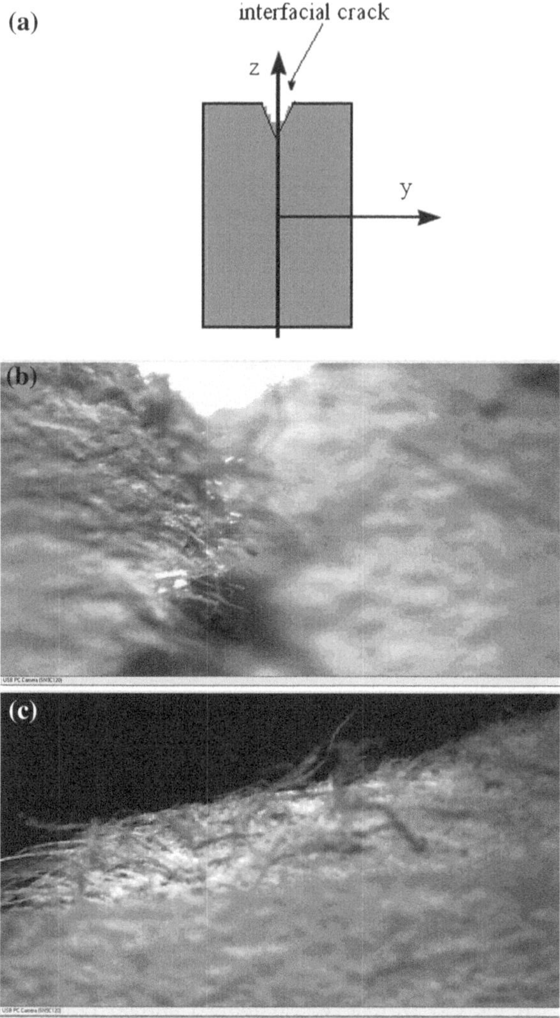

Fig. 3 **a** Theoretical model of crack opening, **b** opened crack in paper (mode I), **c** surface of the paper cut as simulation of environmental macro-world influence on the material

The results of the first experiment are shown in Fig. 3. Simple mode I crack opening is observed.

Corrosion of the metal conductor causes the formation of an insulator or a semiconductor layer, which stops the current flow. Therefore, the corrosion process increases electrical resistance.

Figure 4 shows corrosion under optical microscope. Two magnifications are chosen for the presentation. It can be seen that there are no significant optical differences when using specified magnifications.

Figure 5 shows a compact disk (CD), which have recorded data on the surface. Mechanical damage to CD's surface leads to the loss of data due to the unreadability of the disk.

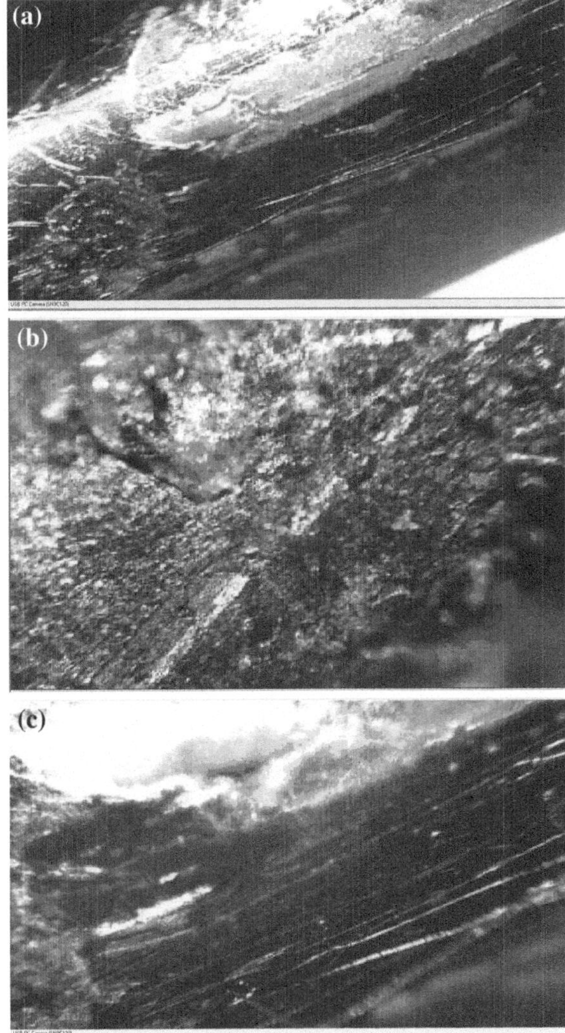

Fig. 4 Corroded screw-driver (metal material sample): **a** 4x magnification, **b**, **c** 10x magnification

Figure 5 is a good example of the impact of magnification on the understanding of the image. Namely, 40x magnification is less understandable to people than 10x magnification of the CD's surface.

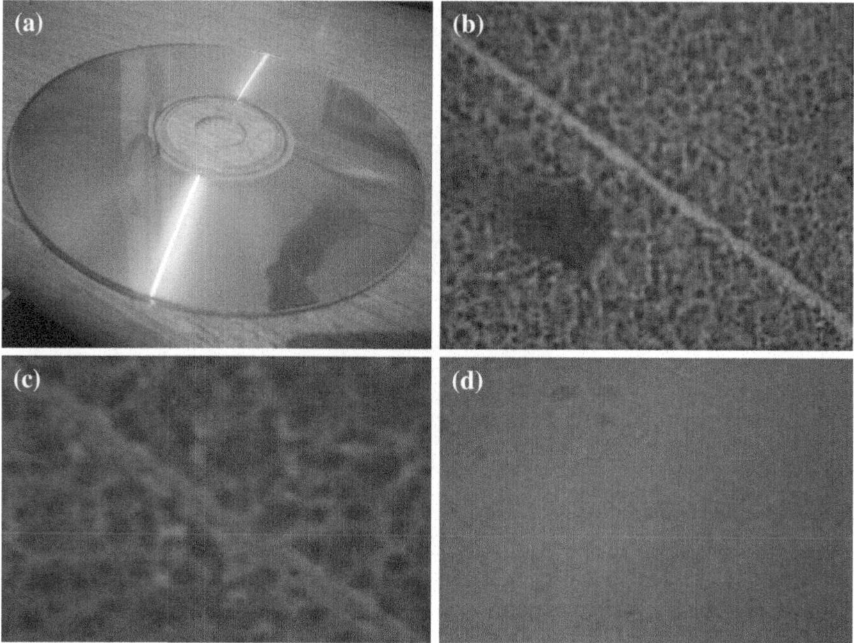

Fig. 5 **a** Recorded CD, **b** CD's surface magnified by 4x, **c** CD's surface magnified by 10x, **d** CD's surface magnified by 40x

4 Conclusion

Integrating the awareness of the degradation of materials is of great importance for the preparation of students for the performance of maintenance duties or designing of technical products. Three examples of mechanical causes of electrical failure are shown in the chapter. Examples are electrical insulator (paper), conductor (metal) and optical storage (CD's surface material, such as $Sb_yGe_{1-y}O_x$). Students learned about them both in theory and by experiment. The theory and experiment are connected by showing the theoretical crack opening and observation of the crack in mode I by optical microscope. This teaching method increased the students' interest in the subject of the course.

References

1. National Instruments Catalog. http://www.ni.com. Accessed 12 March 2011
2. Hornyak, G.L., Tibbals, H.F., Dutta, J., Moore, J.J.: Introduction to Nanoscience & Nanotechnology. CRC Press, Boca Raton (2009)
3. Kuzmanić, I., Vlašić, R., Vujović, I.: Materials in Electrical Engineering. University of Split, Split College of Maritime Studies, Split (2001) (in Croatian)

4. Xun, X.Y., Zhen, X.F., Xi, D.M.: Evaluation of thermal degradation induced material damage using nonlinear lamb waves. Chin. Phys. Lett. (2010). doi:10.1088/0256-307X/27/1/016202
5. Basaran, C., Tang, H., Nie, S.: Experimental damage mechanics of microelectronic solder joints under fatigue loading. In: Proceedings of the 55th electronic components and technology conference, vol. 1, pp. 874–881, 31 May–3 June 2005
6. Courtney, T.H.: Mechanical Behaviour of Materials. Waveland Press, Long Grove (2000)
7. Callister, W.D.: Materials Science and Engineering—An Introduction. Wiley, New York (2007)
8. Vujović, I., Kulenović, Z., Kuzmanić, I., Kezić, D.: Stress intensity factor, energy distribution and forces acting on crack in piezoelectric robots' sensors. In: Katalinic, B. (ed.) DAAAM International Scientific Book 2006. DAAAM International, Vienna (2006)
9. Kulenović, Z., Kuzmanić, I., Vujović, I.: About a crack in piezoceramic element of electromechanical device on the ship. Naše more **52**, 75–80 (2005)
10. Hedrih, K., Lj, Perić: Stanje napona i stanje deformacija u piezoelektričnom materijalu u okolini vrha prsline za slučaj ravne deformacije (in Serbian). Facta Universitatis Mach. Eng. **45**, 50–56 (1996)
11. Hellen, K.: Introduction to Fracture Mechanics. McGraw-Hill, New York (1984)
12. Hummel, R.E.: Electrical Properties of Materials. Springer, New York (2005)
13. Kasap, S.O.: Principles of Electronic Materials and Devices. McGraw Hill, New York (2006)
14. Vujović, I., Kuzmanić, I.: Marine Electrical Engineering and Electronics and Basic Circuits —Outline with Instructions for Lab. Faculty of Maritime Studies, Split (2008) (in Croatian)

Pre-Processing for Image Sequence Visualization Robust to Illumination Variations

Ivica Kuzmanić, Slobodan Marko Beroš, Joško Šoda and Igor Vujović

Abstract Several images (a sequence) may be used to obtain better image quality. This method is perfect for super-resolution algorithms, which improve sub-pixel clarity of the image and allow a more detailed view. It is possible that illumination variations, e.g. those caused by a light source, lessen the benefits of super-resolution algorithms. The reduction of the quantity of such occurrences by stabilizing variations is important. An enhanced stabilization algorithm is proposed for purposes of reduction of variations in illumination. It is based on the energy contained in wavelet coefficients. In the proposed algorithm, energy plays a role of the memory buffer in memory-based techniques of illumination variation reduction. The benefits of the proposed image stabilization are the higher quality of images and better visualization. Possible applications are in surveillance, product quality control, engine monitoring, corrosion monitoring, micro/nano microscopy, etc.

Keywords Illumination variations · Wavelet transform · Super-resolution · Parseval relation · Energy

I. Kuzmanić (✉) · I. Vujović
Faculty of Maritime Studies, University of Split,
Zrinsko-Frankopanska 38,
21000 Split, Croatia
e-mail: ikuzman@pfst.hr

I. Vujović
e-mail: ivujovic@pfst.hr

S. M. Beroš · J. Šoda
Faculty of Electrical Engineering, Mechanical Engineering and Naval Architecture,
University of Split, Ruđera Boškovića bb, 21000 Split, Croatia
e-mail: sberos@fesb.hr

J. Šoda
e-mail: jsoda@fesb.hr

A. Öchsner et al. (eds.), *Design and Analysis of Materials and Engineering Structures*, Advanced Structured Materials 32, DOI: 10.1007/978-3-642-32295-2_4,

1 Introduction

Illumination variations generate additive and multiplicative noise, explored in different applications, such as surveillance [1], face recognition [2, 3] or super-resolution [4]. These variations can disable higher vision applications, such as motion detection, tracking, super-resolution, pattern/object/action recognition, etc. Therefore, more efficient image sequence visualization requires the suppression of illumination.

Super-resolution algorithms are used to obtain higher quality and more details of an image, which is a good starting point for visualization applications. Furthermore, if more details are obtained, more data can be analyzed.

High resolution (HR) image is obtained by analyzing inter-pixel motion in low resolution (LR) image sequence in case of super-resolution applications [5]. In such a case, variations in illumination cause false motion influencing the construction of the HR image.

Super-resolution can be obtained by different algorithms, either non-wavelet e.g. in [6–9] or wavelet [10–12]. The second generation wavelets are more suited for image super-resolution, because more LR frames lead to irregular sampling [13, 14], and due to the sub-pixel displacement between LR frames. However, assumptions on grid (sampling lattice) structure can be made, although this is not necessary [14].

Although variations in illumination are examined in detail in motion detection [1, 15, 16] or face recognition [2, 3], they are rarely taken into account in super-resolution, which is mostly concerned with the changing of spatial coordinates in time [6–14].

This chapter aims to describe a method of suppression of illumination variations in pre-processing in order to improve HR image quality. The proposed pre-processing is based on a wavelet energy model and minimization of illumination variations.

This chapter is organized as follows: an illumination variation model is introduced in the second section. The proposed pre-processing is described in the third section, including the mathematical background. In the forth section, results support the theoretical conclusions by illustrating the improvement in the Signal-to-noise ratio (SNR) and by visual inspection.

2 Illumination Variation Model

Illumination variations occur between any two frames of an image, as illustrated in Fig. 1. In image processing, the image is stored as two- or three-dimensional matrix. Rows and columns are used to address the matrix element of interest. The matrix element value is the value of the image pixel. The number of rows and columns defines the size and resolution of an image.

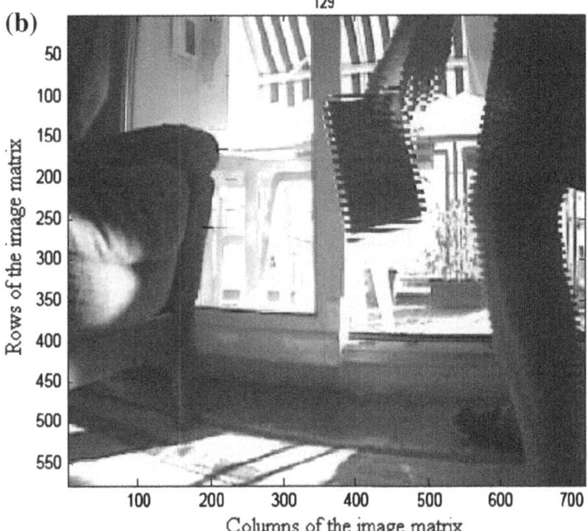

Fig. 1 Illustration of illumination variations—the same scene appears *darker* (*up*) or *lighter* (*down*) under the same lighting conditions (row numbers designate the y-axis and column numbers designate the x-axis; numbers 125 and 129 designate the order of the image in the image sequence)

Illumination variations are visible due to their different effect on brightness. Therefore, Fig. 1a induces the sense of a darker room in the human visual system, while Fig. 1b induces the sense of more light in the scene.

Variations in illumination occur under both natural and artificial light sources. Figure 2 illustrates the measurement for natural and artificial source. Variations in illumination can be positive or negative.

If value of a pixel is increased due to variations in illumination, the variation is called positive. The result of positive variation is increased pixel brightness. Globally speaking, the image looks brighter. However, algorithms can interrupt

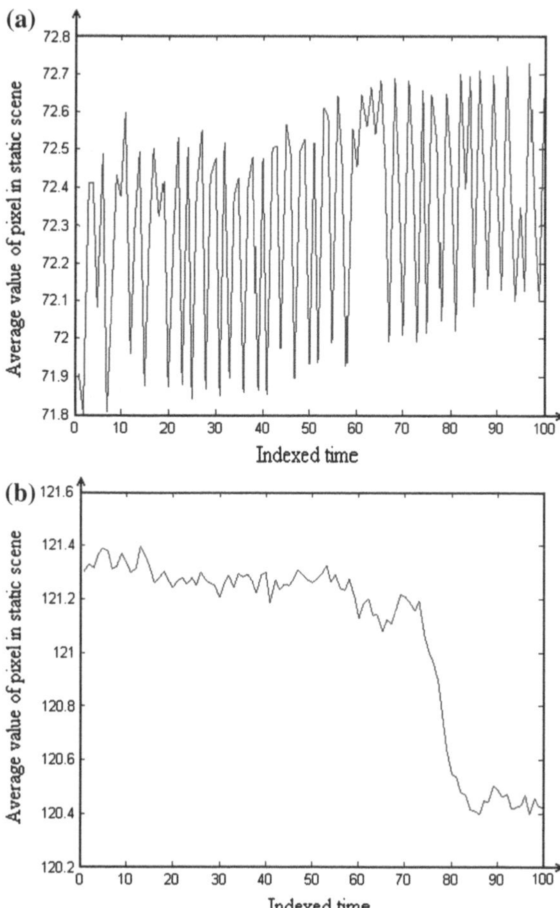

Fig. 2 Experimental change of illumination in the same scene over time: **a** artificial source, **b** natural source

execution due to variations in illumination. For example, if we look at a gray image having the range of 0–255, provided variation in illumination adds 10, all white pixels, originally having the range of 246–255, are out of range. If an algorithm cannot cope with this, the execution will be interrupted. If it is presumed in the programming, then the solution influences the entire image. In some cases, normalized values are used. In such case, the values 0–255 are divided with 255 and the new range is between 0 and 1.

If value of a pixel is decreased due to variations in illumination, the variation is called negative. The result of negative variation is decreased pixel brightness and the image looks darker. In this case, the value of the pixel can be expressed in negative numbers, which have to be predicted in the programming stage of the algorithm. The solution again influences the image perception.

Therefore, it would be useful to suppress or eliminate variations in illumination.

What is the connection between visualization, variations in illumination, image sequences and super-resolution? First, variations in illumination occur in image sequences. Image sequences are used for super-resolution, which can be used for visualization of data, e.g. satellite images. In case of super-resolution, the input is always in low resolution (LR). The goal is always to increase the resolution. Therefore the output is a high resolution (HR) image.

LR images are often blurred, down-sampled, degraded by noise or variations in illumination. Some noise is generated by camera circuitry. This part is covered by the so called camera model [4]. The temporal noise is modelled after additive Gaussian noise. The image gray value level $I(x, y)$ is assumed to be proportionate to the product of reflectance $r(x, y)$ and illumination $e(x, y)$ [2]:

$$I(\mathrm{x},\ \mathrm{y}) = r(\mathrm{x},\ \mathrm{y}) \cdot e(\mathrm{x},\ \mathrm{y}) \tag{1}$$

Of course, Eq. (1) is a way to describe any arbitrary or random noise. There is no way to calculate the actual noise except experimentally, by establishing the differences between supposedly identical images. According to [2], the goal is to recover the reflectance under illumination variations. To simplify the explanation, it is supposed that one of the two neighbouring frames is free of variations in illumination. Additive Gaussian noise is present in both frames. In this case, the two images can be expressed by [1]:

$$I(n) = S(n) + \eta(n) \tag{2}$$

$$I(n+1) = S(n+1) + \eta(n+1) + \xi(n+1) \tag{3}$$

where:

- $I(n)$ is an n-th frame (frame in time t) and $I(n + 1)$ is an $(n + 1)$-th frame (frame in time $t + \Delta T$),
- $S(n)$ and $S(n + 1)$ are projections of the scene onto the image plane,
- $\eta(n)$ and $\eta(n + 1)$ are additive Gaussian noise, and
- $\xi(n + 1)$ is additive and multiplicative illumination variation

Additive and multiplicative variation consist of two components [1]:

$$\xi(n+1) = \mathrm{a} + \mathrm{b} \cdot S(n+1) \tag{4}$$

Constants a and b describe the strength of global changes in illumination. Therefore, image differences can contain artefacts due to illumination variations. Consequently, $(n + i)$th image can be described as:

$$I(n+i) = S(n+i) + \eta(n+i) + \xi(n+i) \tag{5}$$

If all images are supposed to be the same, then:

$$S(n) = S(n+1) = S(n+i)$$

Therefore, summation of all the images gives:

$$\sum_{i=0}^{N-1} I(n+i) = \sum_{i=0}^{N-1} S(n+i) + \sum_{i=0}^{N-1} \eta(n+i) + \sum_{i=0}^{N-1} \xi(n+i) \tag{6}$$

$$\sum_{i=0}^{N-1} S(n+i) = NS(n) \tag{7}$$

$$\sum_{i=0}^{N-1} I(n+i) = NS(n) + \sum_{i=0}^{N-1} \eta(n+i) + \sum_{i=0}^{N-1} \xi(n+i) \tag{8}$$

For large N, illumination variations are averaged to zero:

$$\sum_{i=0}^{N-1} \xi(n+i) = 0 \tag{9}$$

Therefore, illumination variations are eliminated and Eq. (8) becomes:

$$\sum_{i=0}^{N-1} I(n+i) = NS(n) + \sum_{i=0}^{N-1} \eta(n+i) \tag{10}$$

The same can be applied to noise:

$$\sum_{i=0}^{N-1} \eta(n+i),$$

which has zero-mean variance [17]. Consequently, Eq. (10) becomes the clear image of the original scene. This property is used in Wiener filters.

3 Proposed Algorithm

In order to solve the problem of variations in illumination at the input of the super-resolution algorithm, one can start from the noise model. To eliminate the noise, minimization of the mean square error (MSE) can be used. The best filters for the MSE are the Wiener filters. Depending on the application, it is sometimes necessary to reduce the data for processing in order to meet the requirements of the technical system or a request for usage in real-time or limited time.

Another reason to choose another transform is the time–frequency relation. Wavelet transform is the standard tool when signals are non-stationary, i.e. in images with noise.

In order to use the best characteristics of the Wiener filter and the wavelet transform, the algorithm for wavelet energy is derived.

The proposed solution lies in the application of wavelets and the exploration of some properties of the Parseval's theorem in L^2 space [18, 19].

The algorithm used in research is based on the discrete wavelet transform (DWT) to obtain the energy frame descriptors.

3.1 Mathematical Background

Wavelet transform can be implemented in several ways. One of them is filter implementation. In filter implementation, the wavelet transform can be seen as convolution of signal and kernel:

$$y[n] = x[n] * h[n] = h[n] * x[n] \tag{11}$$

where * designates the convolution operation. The Eq. (11) can be written with summation operators:

$$y[n] = \sum_{k=1}^{N} x[k] \cdot h[n-k] = \sum_{k=1}^{N} h[k] \cdot x[n-k] \tag{12}$$

Two-dimensional signals, such as images, need more data to be described. These data are approximation coefficients, as well as horizontal, vertical and diagonal detailed coefficients. Moreover, we must take into account both the frequency-time relation and intensity-position. Therefore, a pair of related functions is used: wavelet and scaling functions. It is possible to calculate one from another. The wavelet decomposition of the image is given with [18, 20]:

$$\begin{aligned} f_{-1}(x,y) = \sum_{n=-\infty}^{\infty} \sum_{p=-\infty}^{\infty} & (a_o(n,p) \cdot \phi(x-n) \cdot \phi(x-n) \\ & + b_o(n,p) \cdot \phi(x-n) \cdot \psi(x-n) \\ & + c_o(n,p) \cdot \psi(x-n) \cdot \phi(x-n) \\ & + d_o(n,p) \cdot \psi(x-n) \cdot \psi(x-n)) \end{aligned} \tag{13}$$

where:

- f is the image,
- a approximation coefficient,
- b, c and d detailed coefficients,
- ψ wavelet function,
- ϕ scaling function,
- x and y place of the pixel in the image

If we define $s_{\phi\phi}(x,y) = \phi(x) \cdot \phi(y)$ and $\{s_{\phi\phi}(x-k, y-l) : k, l \in Z\}$, then [18, 20]:

$$f_o(x,y) = \sum_{i=-\infty}^{\infty} \sum_{j=-\infty}^{\infty} a_o(i,j) \cdot s_{\phi\phi}(x-i, y-j) \tag{14}$$

$$a_o(i,j) = \langle f(x,y), s_{\phi\phi}(x-i, y-j) \rangle \tag{15}$$

$$a_{(j+1)n} = \sum_{k=0}^{N} a_{j,k} \int \phi_{j,k}(t) \cdot \phi_{(j+1),k}(t)dt \tag{16}$$

$$\phi(t) = 2 \sum_{k=0}^{2N-1} h[k] \cdot \phi(2t-k) \tag{17}$$

$$\psi(t) = 2 \sum_{k=0}^{2N-1} g[k] \cdot \phi(2t-k) \tag{18}$$

where h is a low-pass filter and g a high-pass filter.

The implementation of wavelet transform through filter design is common. Alternatively, poly-phase implementation can be used. The proposed algorithm can be used in both filter and poly-phase implementation.

The algorithm is developed as a combination of two applicative requirements:

- detection of low-contrast objects and
- suppression of variations in illumination.

Naturally, since some wavelets are suitable for one purpose and others for another, the idea to combine two mother wavelets with different properties was conceived. In such a combination, each wavelet solves a different problem. The optimum result is obtained by combining the solutions. The problem with combining is the method of its execution. For example, two wavelets do not have the same number of coefficients different than zero. The solution is the utilization of energy, which is a unique descriptor. Therefore, a theorem for combining two energy-based wavelets is arrived at.

Theorem (Calculation of image energy by two separable wavelets) Let F_h and F_g be two-dimensional discrete wavelet transform coefficients of the orthogonal mother wavelets h and g, respectively, obtained by analyzing the image $f(n, m) \in L^2(\mathbf{R})$. If h and g are separable in both dimensions, then energy can be calculated as:

$$\sum_{i=1}^{n} \sum_{j=1}^{m} f^2(i,j) = \frac{1}{nm} \sum_{i=1}^{n} \sum_{j=1}^{m} |F_h(i,j)|^2$$

and

$$\sum_{i=1}^{n} \sum_{j=1}^{m} f^2(i,j) = \frac{1}{nm} \sum_{i=1}^{n} \sum_{j=1}^{m} |F_g(i,j)|^2.$$

In such case it is valid to calculate energy as:

$$\sum_{i=1}^{n}\sum_{j=1}^{m} f^2(i,j) = \frac{1}{2nm}\sum_{i=1}^{n}\sum_{j=1}^{m}\left(|F_h(i,j)|^2 + |F_g(i,j)|^2\right) \tag{19}$$

Proof

$$2\sum_{i=1}^{n}\sum_{j=1}^{m} f^2(i,j) = \frac{1}{nm}\sum_{i=1}^{n}\sum_{j=1}^{m}\left(|F_h(i,j)|^2\right) + \frac{1}{nm}\sum_{i=1}^{n}\sum_{j=1}^{m}\left(|F_g(i,j)|^2\right)$$
$$= \frac{1}{nm}\sum_{i=1}^{n}\sum_{j=1}^{m}\left(|F_h(i,j)|^2 + |F_g(i,j)|^2\right)$$

$$\sum_{i=1}^{n}\sum_{j=1}^{m} f^2(i,j) = \frac{1}{2nm}\sum_{i=1}^{n}\sum_{j=1}^{m}\left(|F_h(i,j)|^2 + |F_g(i,j)|^2\right)$$ □

3.2 Step by Step Description of the Algorithm

The choice of wavelets is based on the solution, which is supposed to be meaningful. The second criterion is the suppression of variations in illumination. The third criterion is the orthogonality of the chosen wavelets (because of the execution speed). Daubechies wavelet with 2 vanishing moments (db2) was chosen to cope with illumination variations and db7 to increase the accuracy of edge localization and detect low-contrast objects.

The steps of the proposed algorithm for pre-filtering are as follows:

1. Initialization of the variables.
2. Pre-filtering of the image sequence by wavelet energy and averaging. This step consists of the following:
   - 2a. Image loading,
   - 2b. 2D-Lifting wavelet transform (2D-LWT) by two different wavelets (i.e. db2 and db7) separately,
   - 2c. Summation of the coefficients (each separately) with the previous sum,
   - 2d. Repeat 2a–2c for 49 times.
   - 2e. Calculation of the coefficient average
   - 2f. Energy calculation
   - 2g. Inverse 2D-LWT

In step 2, which is, actually, proposed pre-filtering, LR image sequence (with N frames) is transformed by 2D-LWT using two different wavelets (marked as wav1 and wav2 in expressions). 2D-LWT is performed on each frame separately. Then, the average of all results is taken:

$$pf_1 = \frac{1}{N}\sum_{i=1}^{N} coef_i^{wav1} \tag{20}$$

$$pf_2 = \frac{1}{N}\sum_{i=1}^{N} coef_i^{wav2} \tag{21}$$

To obtain the energy equivalence, Eq. (22) is performed:

$$pf = \frac{pf_1^2 + pf_2^2}{\max(pf_1^2 + pf_2^2)} \tag{22}$$

3. Execution of the algorithm, which has to be enhanced (the simplest are the zoom, linear interpolation or the motion field). The input to this algorithm is the product of step 2.

4 Results

Methods of interpolation are not the scope of the chapter and are so simple just to illustrate the improvement by the proposed pre-filtering. Common measure of the image quality is signal-to-noise ratio (SNR), which is a measure for the contrast between the time-averaged background and the foreground object compared to the total noise. SNR is defined as [17]:

$$SNR = 20\log_{10}\left(\frac{\frac{1}{K}\sum_{k=1}^{K}\bar{I}_{foreground}(k) - \frac{1}{K}\sum_{k=1}^{K}\bar{I}_{background}(k)}{\sigma_n}\right) \tag{23}$$

where:

- K is the number of frames,
- $\bar{I}_{foreground}(k)$ the average foreground intensity in frame k,
- $\bar{I}_{background}(k)$the average background intensity in frame k, and
- σ is the standard deviation of K frames

Methods for super-resolution with higher signal-to-noise ratio will also be improved, expectedly, by the same factor (about 3 times) as the methods in the example.

Table 1 shows the SNR (defined by the expression in [17, 21]) improvement of different algorithms by the proposed pre-filtering. The analyzed images and the obtained results are obtained for this research.

Table 1 Comparison of the SNR for the same methods with and without applying proposed prefiltering

METHOD	SNR [dB]
Original image	3.2
Upsampled and linearly interpolated	3.2177
Upsampled and linearly interpolated with wavelet energy (RMS) and scaling (proposed prefiltering)	10.0658
Mean and motion field interpolation	10.046
Mean and motion field interpolation with wavelet energy (RMS) and scaling (proposed prefiltering)	27.1387
Mean and motion field interpolation with wavelet energy (normalized energy) and scaling (proposed prefiltering)	49.325

Table 2 Low-resolution grid with four elements

LR (1,1)	LR (1,2)
LR (2,1)	LR (2,2)

Table 3 Location of low-resolution matrix elements (from Table 2.) on high-resolution grid and expressions of approximations (for every coefficient) [22]

LR (1,1)	A	LR (1,2)
C	D	E
LR (2,1)	H	LR (2,2)

Explanation of the letters:
$A = \frac{1}{2}[LR(1,1) + LR(1,2)]$
$C = \frac{1}{2}[LR(1,1) + LR(2,1)]$
$D = \frac{1}{2}\left[LR(1,1) + \frac{1}{2}(LR(1,2) + LR(2,1))\right]$
$E = \frac{1}{2}[LR(1,2) + LR(2,2)]$
$H = \frac{1}{2}[LR(2,1) + LR(2,2)]$

As can be seen from Table 1, the original signal has SNR equal to 3.2 dB. The usual zoom does not increase SNR value.

However, if the proposed pre-filtering is applied, then SNR is increased by the magnification factor (*MF*):

$$MF = \frac{10}{3.2} = 3.125$$

MF = 3.125 means that the SNR is increased by 3.125 times.

However, if the motion field is used to increase the number of pixels (resolution) then the obtained image without proposed pre-filtering results in SNR about 10 dB. If this method is improved as proposed, the SNR rises to 27 dB.

Simple linear interpolation is performed by calculating the averages between low-resolution pixels (Table 2) in the high-resolution grid [22]. Low-resolution pixels position in the high-resolution grid is shown in Table 3.

(a)
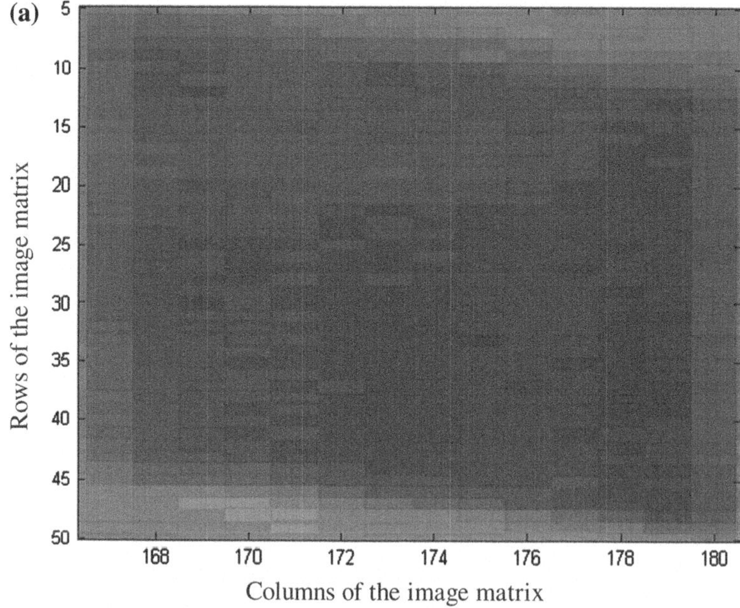

(b)

Fig. 3 **a** Zoomed part of the original image, **b** original image, **c** zoomed part of the image obtained by simple super-resolution algorithm, **d** the entire image obtained by the super-resolution

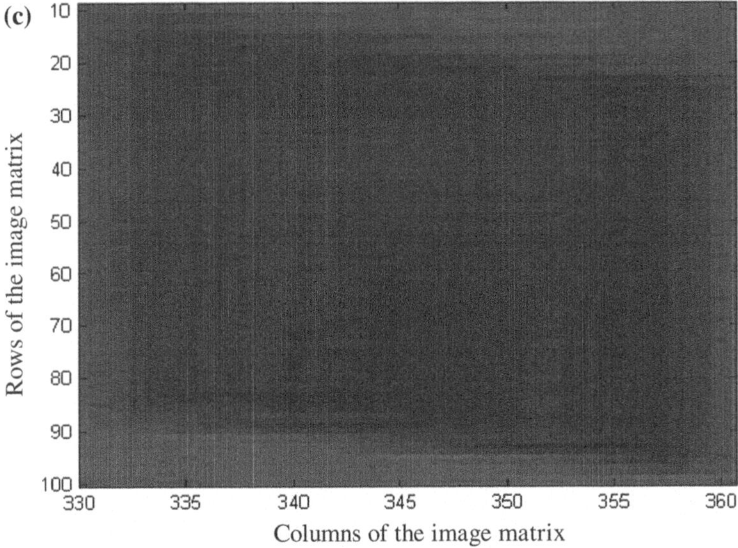

Fig. 3 (continued)

The problem of the zoom is the visible grain structure, as can be seen in Fig. 3. Figure 3a shows the zoomed part of the source and Fig. 3b the entire source.

Figure 3c shows the same zoomed part as Fig. 3a, but the image is improved by super-resolution algorithm. Hence, the number of rows and columns for the same image part is doubled.

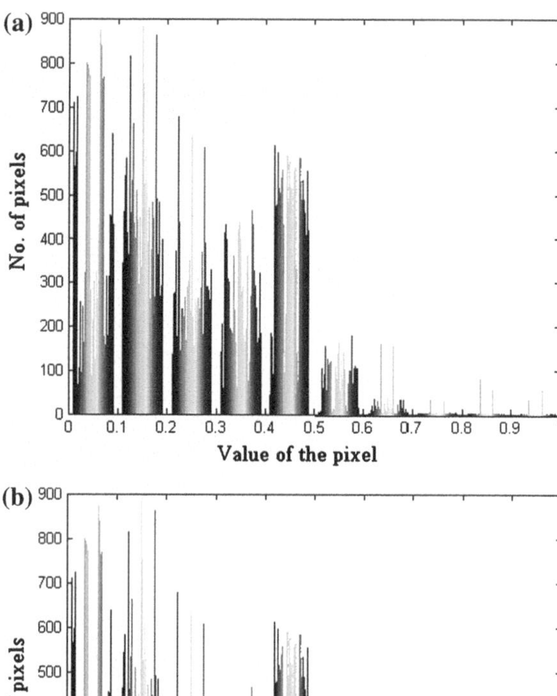

Fig. 4 Histogram of: **a** the original image, **b** the image obtained by wavelet quasi-super-resolution [23]

Figure 3d shows the entire image obtained by the implementation of the super-resolution algorithm.

In order to validate the proposed algorithm, other algorithms from references should be used for comparison.

Statistical mean to evaluate the concentration of the interesting value in the set of data is called histogram.

Histograms in Fig. 4 show number of pixels with the same value and the distribution through the entire value range. The comparison of the histograms in the quasi-super-resolution algorithm [23] and the original image is shown. It can be seen that histograms are almost identical, which means that it is very likely that human visual perception remained unchanged with respect to the addressed problem, assuming that it is the same scene, within the same context, shot under the same weather conditions (if it is in outdoor application), at the same time of day. Of course, in [23] a method is presented, which enhanced the usual zoom operation with wavelets and interpolation.

Super-resolution result obtained by linear interpolation, super-resolution result obtained by linear interpolation and the proposed prefiltering and the image difference between two results are shown in Fig. 5. The numbers that designate the axes are number of rows and columns, which means the adress of the pixel. I.e. (3,5) means that we seek for the pixel in the 3rd row and 5th column.

An image difference is defined as [4]:

$$\delta_{k,m} = y_{k,m} - \sum_{r=1}^{R} w_{k,m,r} \tilde{b}_{k,r} \tag{24}$$

where:

- $w_{k,m,r}$ represents the operations of blurring and down-sampling,
- $\tilde{b}_{k,r}$ is shifted HR r-th pixel in k-th frame, and
- $y_{k,m}$ is the source (LR-image) pixe

In order to calculate the difference, the automated algorithm or human operator can be used. The disadvantage of the automated algorithm is the inherit error due to decision threshold. The disadvantage of the human operator is the speed of execution. Using the second choice, which is more precise and more time consuming, it is calculated that Fig. 5c consists of:

- 8836 pixels which are exactly the same (0.1775 %),
- 1372834 pixels are near zero or equal to zero, which means that the images, which are subtracting, are almost the same (27.5797 %) and
- 3596033 pixels are different (72.2428 %).

Pixels that are different present the difference between two methods: the standard linear interpolation and the linear interpolation, which is improved by the proposed prefiltering.

It can be observed that Fig. 5b also has less additive noise.

Figure 6a shows an example of the source image obtained by the frame grabbing procedure form the camera. It means that it is only one of the frames from the continuous image sequence in on-line operation mode.

It is important to evaluate different algorithms in the same situations, i.e. the same scene or the same time and the same scene. Otherwise the results can be totally different and force us to arrive at a different conclusion. Therefore, Fig. 6 shows the same sourced image and the zoomed part. The zoomed part is at the edge of armchair.

Figure 7 presents the results of different methods of interpolation in the super-resolution algorithm to the same source image.

Figure 7a shows the resulting image obtained by simple linear interpolation and up-sampling operation. If the proposed pre-processing algorithm is applied to the same procedure, the obtained result is shown in Fig. 7b.

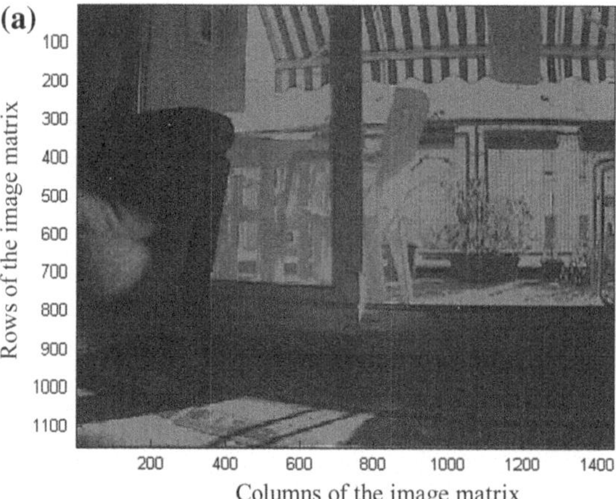

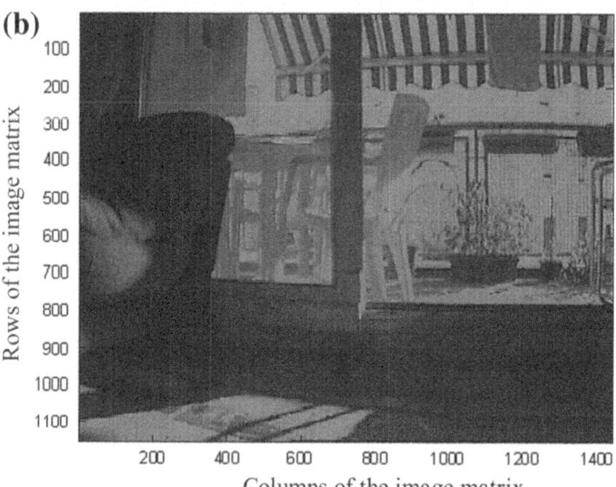

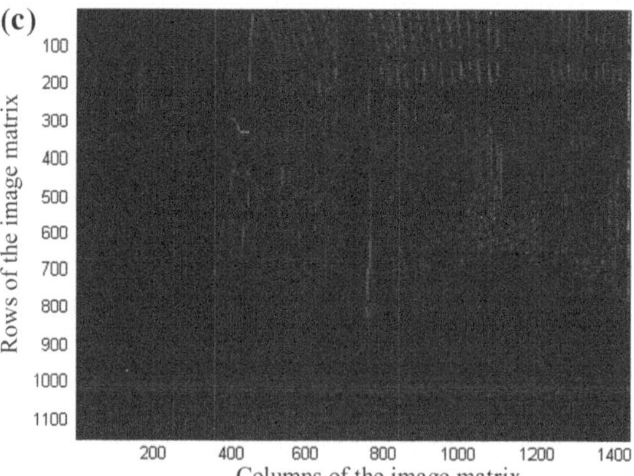

Fig. 5 **a** Super-resolution by linear interpolation, **b** super-resolution by linear interpolation with the proposed pre-filtering, **c** the image difference between **a** and **b**

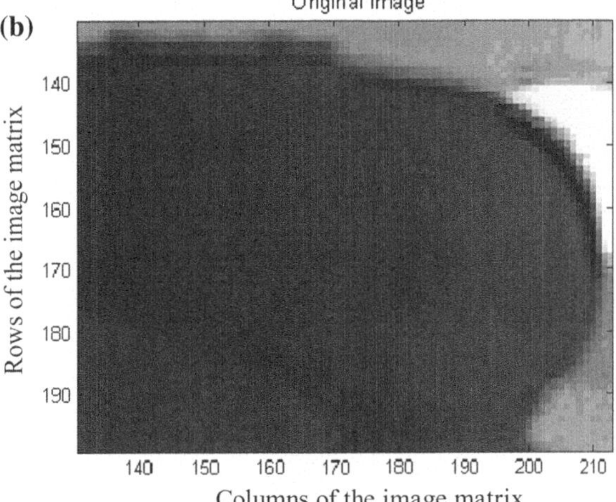

Fig. 6 Original image (**a**) and zoomed part of the same image (**b**)

The experimental scene has different interesting moments for analysis. One of them is the edge of the armchair. Figure 7c shows that interesting detail of the scene, which is placed between 275 and 390th row and 390 and 420th column.

The motion field method examines the interframe motion in the image sequence. Figure 7d shows the result of such simple algorithm. Zoomed part (between 260 and 390th row and 280 and 425th column) of the same region of interest (ROI) is shown in Fig. 7e. This part is magnified from Fig. 7c.

Figure 7f is an image of energies for the pixels corresponding to the scene.

If the proposed pre-processing algorithm is applied to the image and then the motion field method, the result is an improved motion-field method. The result is shown in Fig. 7g. For the sake of comparison, the same ROI (between 260 and 390th row and 280 and 425th column) as in other methods is zoomed and shown for this method in Fig. 7h.

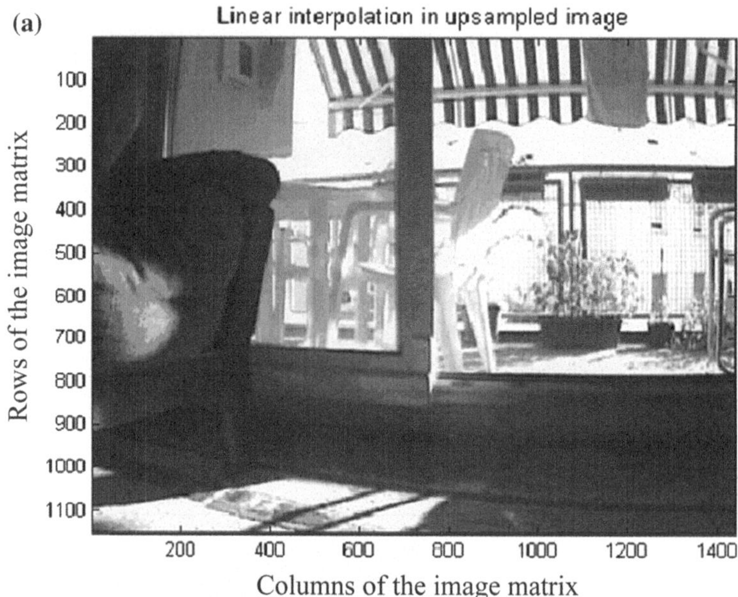

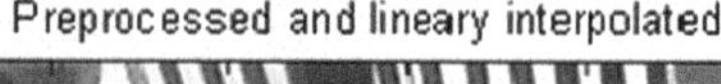

Fig. 7 Different methods of interpolation in super-resolution: **a** result of linear interpolation and up-sampling, **b** the same as **a** with proposed pre-filtering, **c** zoomed part of **a** between 275 and 390th row and 390 and 420th column, **d** motion field interpolation with pre-processing as proposed, **e** zoomed part of **b** between 260 and 390th row and 280 and 425th column, **f** energy model of the scene, **g** motion field method with the proposed pre-processing in energy, **h** zoomed part of **g** between 260 and 390th row and 280 and 425th column, **i** high resolution image obtained by energy model, **j** root-mean square method with linear interpolation, **k** zoomed part of **j** with scaling factor π between 250 and 390th row and 280 and 425th column, **l** absolute difference between previous two methods of super-resolution, **m** image **l** with scaling factor π

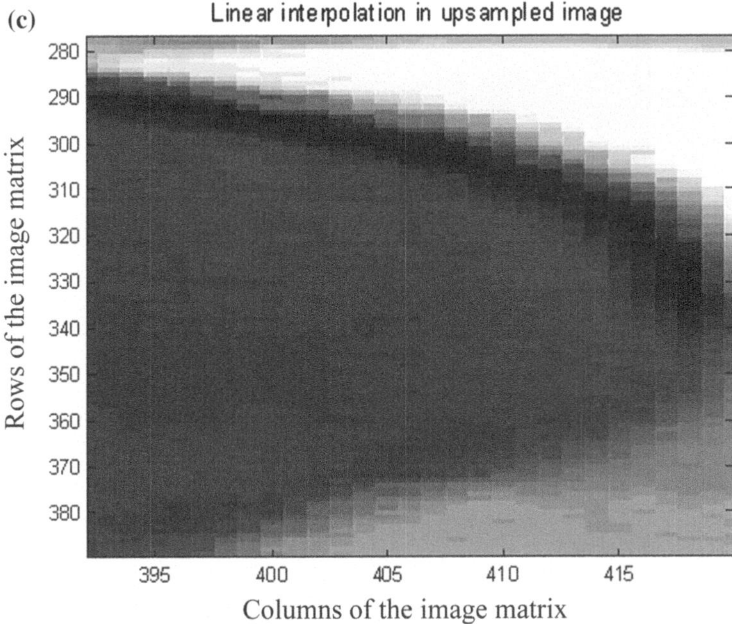

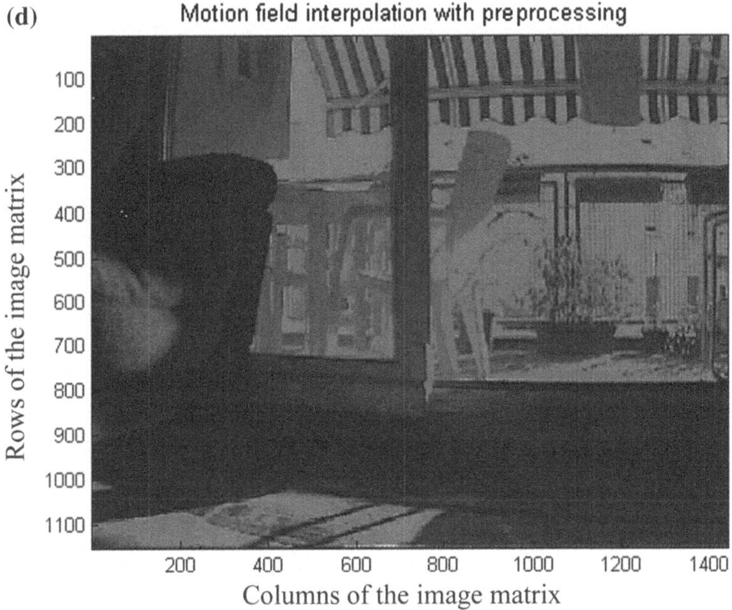

Fig. 7 (continued)

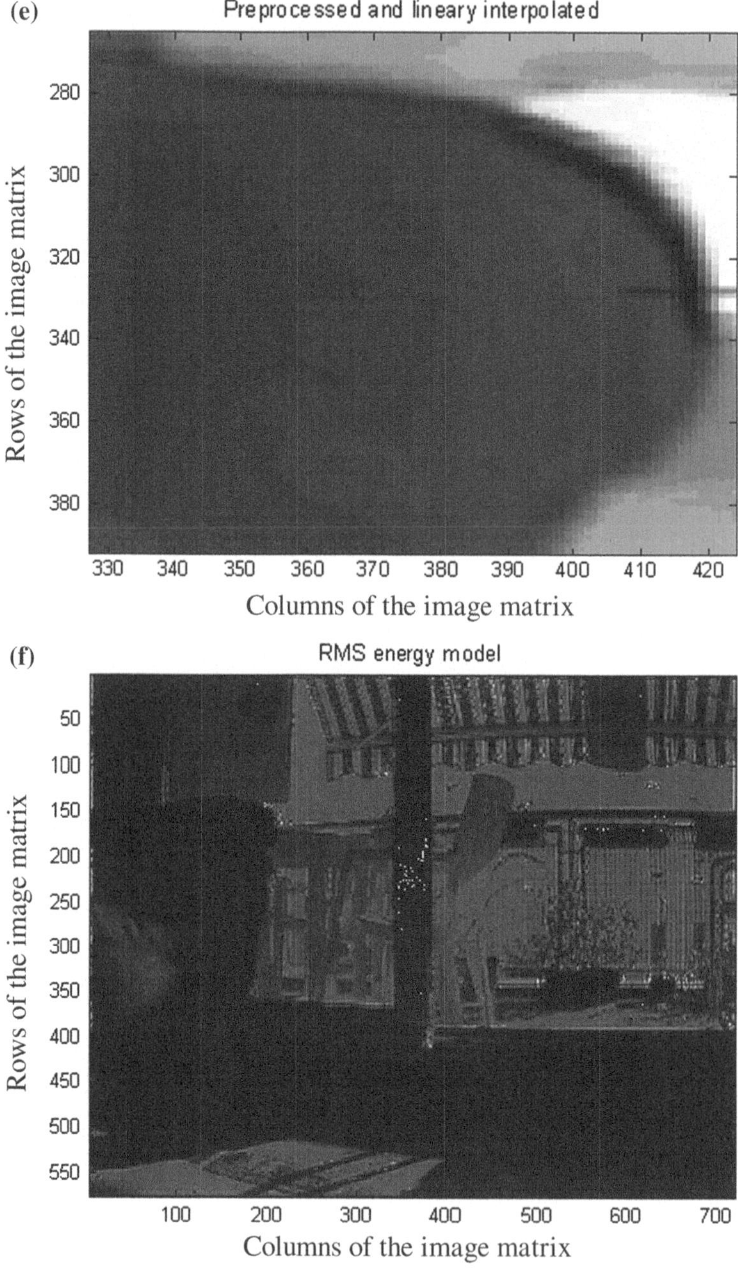

Fig. 7 (continued)

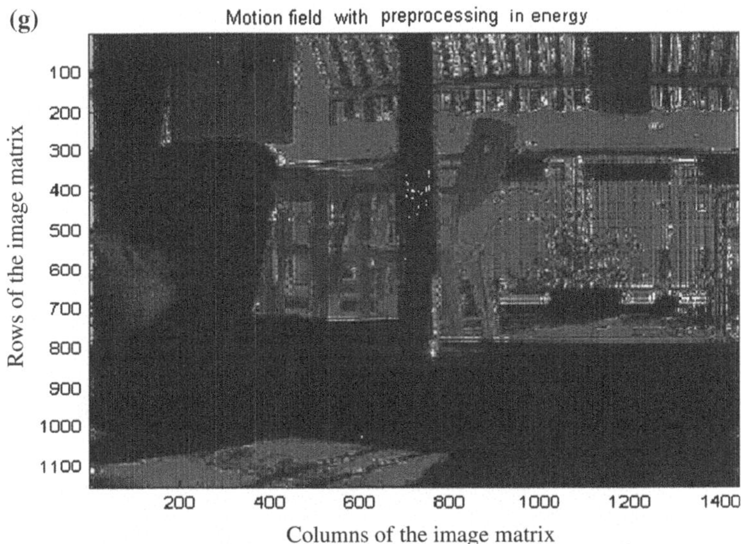

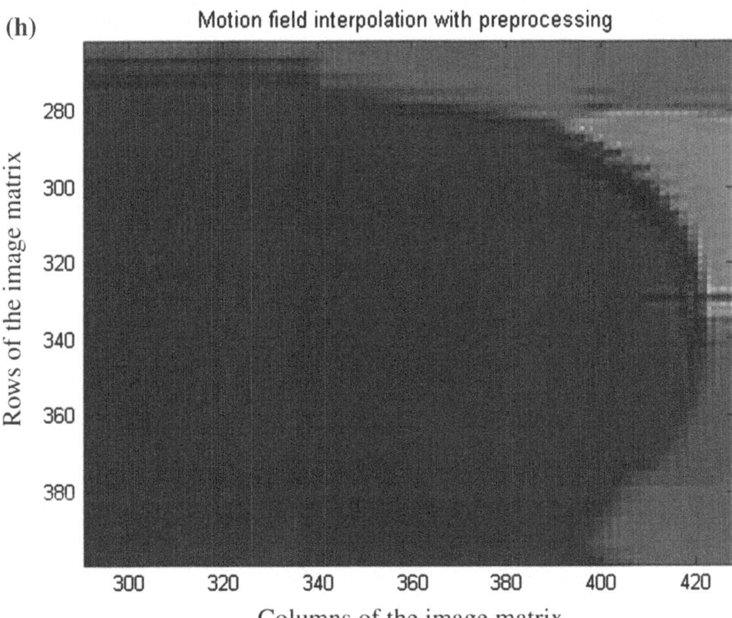

Fig. 7 (continued)

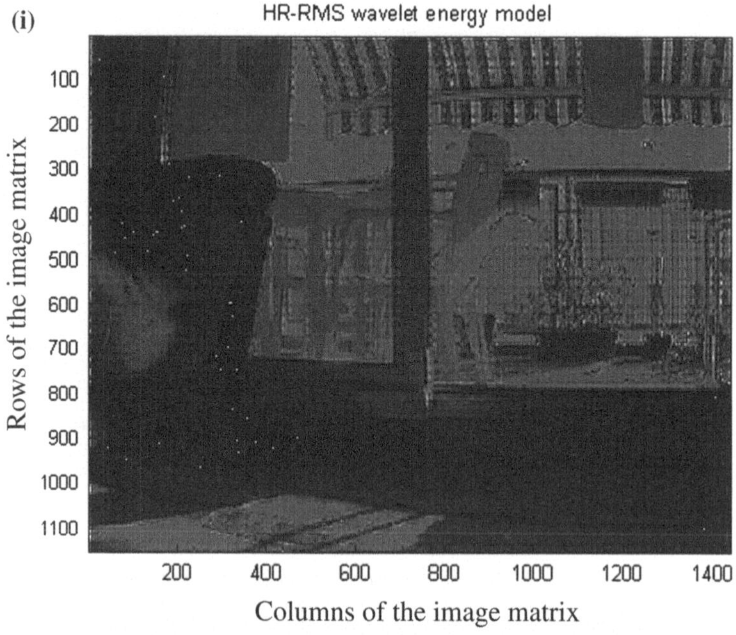

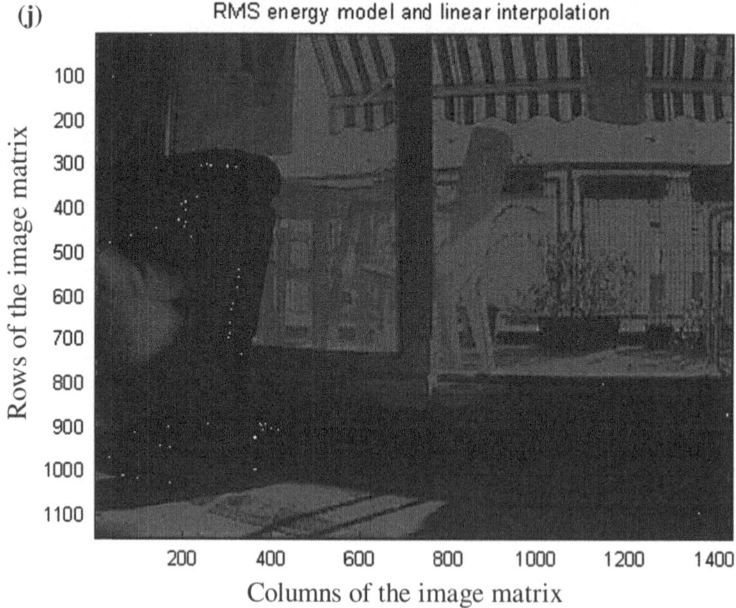

Fig. 7 (continued)

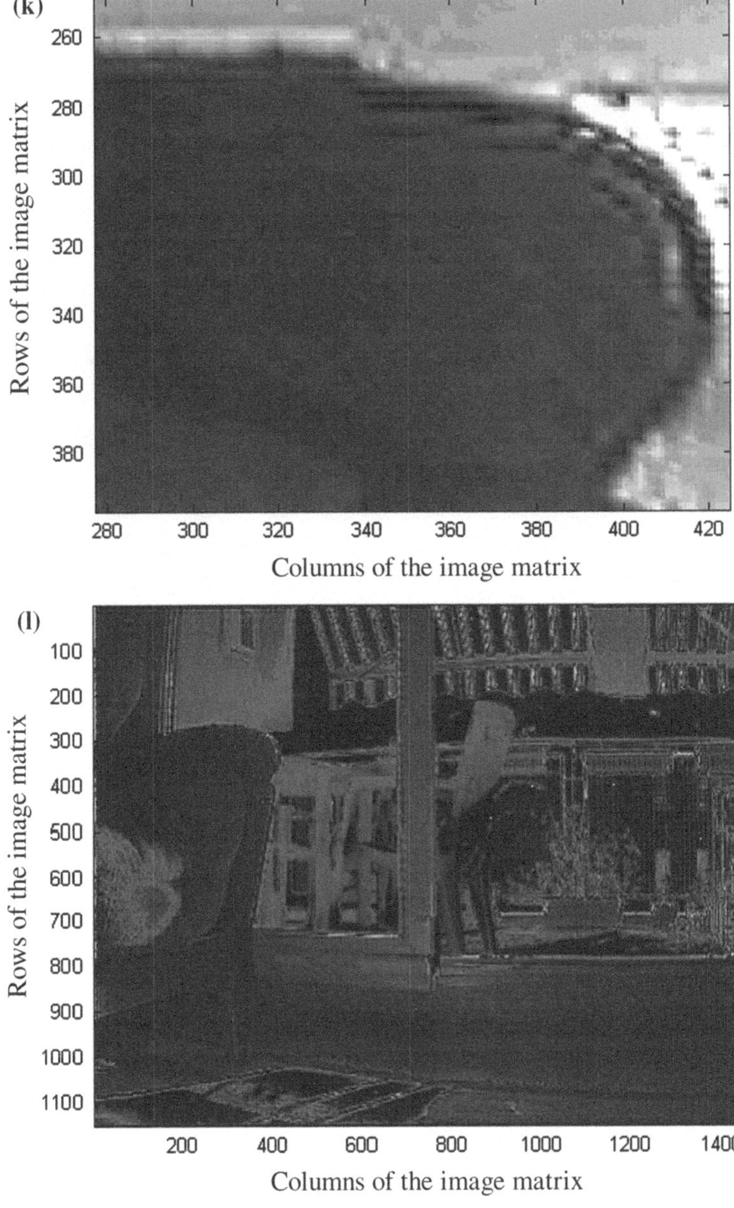

Fig. 7 (continued)

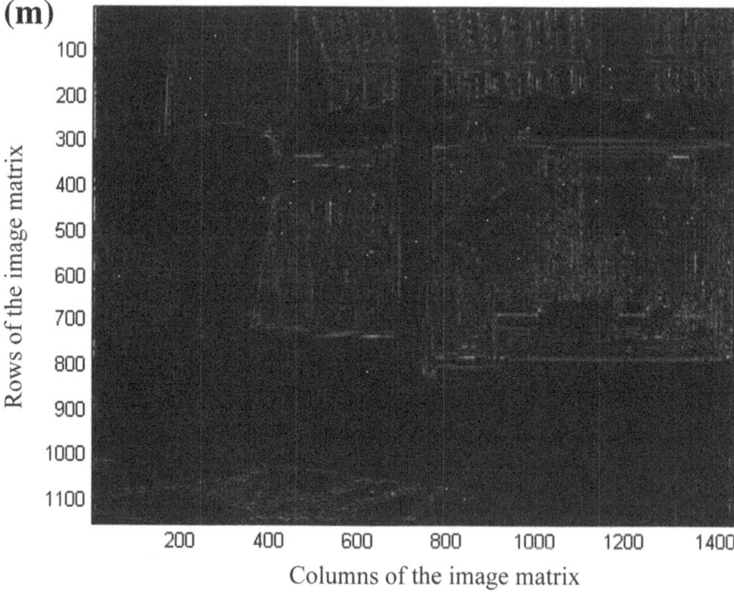

Fig. 7 (continued)

If the energy method is used, the obtained high resolution (HR) image is shown in Fig. 7i.

The result of root-mean square (RMS) method applied together with linear interpolation is shown in Fig. 7j. The same ROI (between 260 and 390th row and 280 and 425th column) as for the other methods is used to obtain the Fig. 7k. The final result is scaled with factor π.

The absolute difference between the two presented super-resolution methods is shown in Fig. 7l.

If scaling operation is applied to image difference and the scaling factor is π, the result is shown in Fig. 7m.

Figures 6 and 7 illustrate the original image, zoomed part of the image, and super-resolved image with different methods. However, there are also many other methods. The chosen methods are for illustration purposes only. The proposed algorithm can also be used to improve any other method.

5 Conclusion

The scope of this chapter is the improvement of input data of the super-resolution visualization algorithm. The method based on energy defined in the wavelet domain is proposed and root-mean square is used to minimize variations in illumination. The proposed pre-filtering method improves the discussed algorithms by a factor of 3, if

SNR is used for comparison. The SNR improvement is the result of the proposed algorithm. However, the results are obtained by root-mean square minimization of variations in the LR image sequence. This procedure introduces more details in the sub-pixel level. The proposed algorithm can be applied in various visual systems, such as surveillance (motion detection), product quality control (pattern matching or recognition), engine monitoring, corrosion monitoring (pattern recognition), micro/nano microscopy (for the educational or research purposes), etc.

References

1. Amer, A.: Memory-based spatio-temporal real-time object segmentation for video surveillance. In: Proceedings of the Conference on Real-time Imaging VII, Santa Clara, CA, vol. 5012, pp. 10–21. 22–23 Jan 2003
2. Zhichao, L., Joo, E.M.: Face recognition under varying illumination. In: Er, M.J. (ed.) New Trends in Technologies: Control, Management, Computational Intelligence and Network Systems, InTech, Rijeka (2010)
3. Perronnin, F., Dugelay, J.L.: A model of illumination variation for robust face recognition. Workshop on multimodal user authentication, Santa Barbara, USA, 11–12 Dec 2003
4. Eekeren, A.W.M., Schutte, K., Vliet, L.J.: Multiframe super-resolution reconstruction of small moving objects. IEEE Trans. Image Process **19**, 2901–2912 (2010)
5. Robinson, M.D., Toth, C.A., Lo, J.Y., Farsiu, S.: Efficient fourier-wavelet super-resolution. IEEE Trans. Image Process **19**, 2669–2681 (2010)
6. He, Y., Yap, K.H., Chen, L., Chau, L.P.: A nonlinear least square technique for simultaneous image registration and super-resolution. IEEE Trans. Image Process **16**, 2830–2841 (2007)
7. Brito, A.E., Chan, S.H., Cabrera, S.D.: SAR image superresolution via 2-D adaptive extrapolation. Multidimension. Syst. Signal Process. **14**, 83–104 (2003)
8. Ng, M.K., Yau, A.C.: Super-resolution image restoration from blurred low-resolution images. J Math Imaging Vis **23**, 367–378 (2005)
9. Vandewalle, P.: Super-resolution from unregistered aliased images. Ph.D. thesis, École Polytechnique Fédérale De Lausanne (2006)
10. Nguyen, N., Milanfar, P.: A wavelet-based interpolation-restoration method for superresolution (wavelet superresolution). Circ. Syst. Signal Process. **19**, 321–338 (2000)
11. Bose, N.K.: Image phase-only information for landmine classification using ANN and DT/wavelet superresolution from image sequence. Sixth Annual Army Landmine Research Technical Review Meeting, Springfield, VA, 23 Jan 2003
12. Mastriani, M.: New wavelet-based superresolution algorithm for speckle reduction in SAR images. Int. J. Comp. Sci. **1**, 291–298 (2006)
13. Bose, N.K., Letrattanapanich, S., Chappalli, M.B.: Superresolution with second generation wavelets. Signal Process. Image **19**, 387–391 (2004)
14. Bose, N.K., Chappalli, M.B.: A second-generation wavelet framework for super-resolution with noise filtering. Int. J. Imaging Syst. Technol. **14**, 84–89 (2004)
15. Rosin, P., Ioannidis, E.: Evaluation of global image thresholding for change detection. Pattern Recogn. Lett. **24**, 2345–2356 (2003)
16. Porter, R., Fraser, A.M., Hush, D.: Wide-area motion imagery. IEEE Signal Process. Mag. **27**, 56–65 (2010)
17. Dorf, R.C.: The Electrical Engineering Handbook. CRC Press LLC, Boca Raton (2000)
18. Mallat, S.: A Wavelet Tour of Signal Processing, 2nd edn. edn. Academic Press, New York (1999)
19. Poularikas, A.D.: Signals and Systems Primer with Matlab. CRC Press, New York (2007)

20. Mertins, A.: Signal Analysis: Wavelets, Filter Banks Time-Frequency Transforms and Applications. Wiley, West Sussex (1999)
21. Eekeren, A.W.M., Schutte, K., Vliet, L.J.: Multiframe Super-Resolution Reconstruction of Small Moving Objects. IEEE Trans. Image Process. **19**, 2901–2912 (2010)
22. Vujović, I., Kuzmanić, I., Vujović, M.: Algorithm for combined wavelet quasi-superresolution. In: Proceedings of 5th International Symposium Communication Systems Networks and Digital Signal Processing, Patras, Greece, vol. 1, pp. 469–473, 19–21 July 2006
23. Vujović, I., Kuzmanić, I.: Wavelet quasi-superresolution in marine applications. In: Proceedings of the 48th International Symposium ELMAR—2006 focused on Multimedia Signal Processing and Communications, Zadar, Croatia, vol. 1, pp. 65–68 (2006)

Efficient Crack Propagation Simulation Using the Superimposed Finite Element Method and Cohesive Zone Model

Y. T. Kim, H. C. Oh and B. C. Lee

Abstract There have been studies for the crack propagation simulation such as the node release technique, the element elimination method, the cohesive zone model, the extended finite element method, etc. Among these methods, the cohesive zone model is known as an effective method for the crack initiation and propagation. The cohesive zone model is easy to implement in a finite element code and estimates accurately the experimental results. Unfortunately, it has some drawbacks. The crack path is already known for the crack propagation because the cohesive elements are already generated on the crack path. Additionally, it is difficult to generate the cohesive element because the thickness of the cohesive element is very thin. In this study, an effective method by using the superimposed finite element method is proposed to overcome these drawbacks. The superimposed finite element method is one of the local mesh refinement methods. A fine mesh is generated by overlaying the patch of the local mesh on the existing mesh called the global mesh. Thus, re-meshing is not required. When the crack propagates, the local mesh refinement by using the superimposed finite element method is operated using the local element patch. The mesh of the local element patch includes the crack and the cohesive elements are generated on the crack surface of the local element patch. Therefore, the crack propagation simulation can be performed along the new crack path. Also, some local mesh patches are classified as the direction of the crack propagation in the proposed method and the local mesh

Y. T. Kim (✉) · H. C. Oh · B. C. Lee
Korea Advanced Institute of Science and Technology,
Yu-seonggu, Daejeon, South Korea
e-mail: Ytaekim@gmail.com

H. C. Oh
e-mail: hcoh@kaist.ac.kr

B. C. Lee
e-mail: bclee@kaist.ac.kr

A. Öchsner et al. (eds.), *Design and Analysis of Materials and Engineering Structures*, Advanced Structured Materials 32, DOI: 10.1007/978-3-642-32295-2_5,

patch is generated using the hierarchical concept. Then, the generation of the local mesh patch is easy and efficient. Additionally, the re-meshing process is not required. Consequently, the proposed method improves the efficiency of the crack propagation simulation. The proposed method is applied to several examples.

1 Introduction

During the past decades, methods for the crack propagation simulation have been studied actively. Typical methods are the node release technique, the element elimination method, the extended finite element method and the cohesive zone model, etc. The node release technique [1] is the method of separating two nodes directly when a certain value such as the stress or the strain reaches a critical value at that node. The element elimination method [2] is the method of removing elements which satisfy some failure condition. When the failure condition is satisfied, the material stiffness is set to be null. This method can be applied when the crack path is not prescribed. The extended finite element method [3–6] uses elements with embedded discontinuities. In the extended finite element method, discrete crack propagation through a finite element mesh is modeled by the enrichment of the classical displacement-based finite element approximations by a discontinuous function obtained through the partition of unity. The cohesive zone model [7, 8], which is one of the interface modeling methods, was introduced to model the interface. The complicated behavior of the interface is characterized into one cohesive layer. The material behavior of the interface is determined by a traction-separation law.

The node release technique and the element elimination method are simple but they do not have the capability to simulate a true discontinuity. Furthermore, the extended finite element method has some difficulties which lie in the modeling of spontaneous multiple crack initiation, branching and coalescence and the complicated numerical integration.

On the other hand, the cohesive zone model is easy to implement in a finite element code and estimates accurately the experimental results. Unfortunately, it has some drawbacks. For using the cohesive zone model, the cohesive elements are generated on the crack surface after the crack path is determined during the analysis. So, re-meshing process is required repeatedly. Additionally, it is difficult to generate the cohesive element because the thickness of the cohesive element is very thin.

In this study, an effective method by using the superimposed finite element method is proposed to overcome these drawbacks. As one of the mesh refinement techniques, the superimposed finite element method was proposed by Fish [9]. The idea of this method is to overlay a fine local mesh into the concerned area, which is discretized as a rough global mesh. A re-meshing process is not needed with the superimposed finite element method, and the performance is similar to a fine mesh.

Due to this advantage, this method is applied to various fields such as laminated composites [10], elasto-dynamic problems [11], crack propagation [12], shape optimization [13], simulation of the adhesive joint [14], etc. Recently, an efficient superimposed finite element method has been proposed by Park et al. [15]. As the boundary in the global mesh coincides with the boundary in the local mesh, computational cost is reduced. This method is applied to the crack propagation simulation. When the crack propagates, the local mesh included the crack is generated. This mesh is superimposed on the global elements which include the crack. The local mesh includes the cohesive elements on the crack surface. Therefore, the crack propagation simulation can be performed with the cohesive elements.

For the cohesive elements are generated easy, the divided patterns of the global element are defined. Using these patterns, the local mesh patch is generated with the hierarchical concept. Then, the generation of the local mesh patch is easy and efficient. Furthermore, the mesh density control of the local fine element is also easy. Additionally, the re-meshing process is not required. This is the most important fact of the superimposed finite element method. Consequently, the proposed method improves the efficiency of the crack propagation simulation. The proposed method is applied to several examples.

2 Superimposed Finite Element Method

The governing equation of the finite element method is obtained from the variation of the total potential energy functional Π in Eq. (1).

$$\Pi = \frac{1}{2}\int_{\Omega} \varepsilon^T \mathrm{D} \varepsilon \, d\Omega - \int_{\Gamma} \mathrm{u}^T t \, d\Gamma - \sum_i \mathrm{u}_i^T \mathrm{f}_i \tag{1}$$

where Ω is the domain, ε is the strain, D is the material matrix, t and Γ are the traction boundary condition and the region where the traction is applied, f_i and u_i are the external force and the displacement in the point where the external force is applied. The variation of Eq. (1) gives Eq. (2)

$$\begin{aligned} \delta\Pi &= \int_{\Omega} \delta\varepsilon^T \mathrm{D}\varepsilon \, d\Omega - \int_{\Gamma} \delta\mathrm{u}^T t \, d\Gamma - \sum_i \delta\mathrm{u}_i^T \mathrm{f}_i \\ &= \sum^{N} \int_{\Omega^e} \delta(\varepsilon^e)^T \mathrm{D}\varepsilon^e \, d\Omega^e - \int_{\Gamma} \delta\mathrm{u}^T t \, d\Gamma - \sum_i \delta\mathrm{u}_i^T \mathrm{f}_i = 0 \end{aligned} \tag{2}$$

where Ω^e is the domain of each element, ε^e is the strain of each element, N is the number of elements.

For the superimposed finite element method, the displacement is defined by Eq. (3).

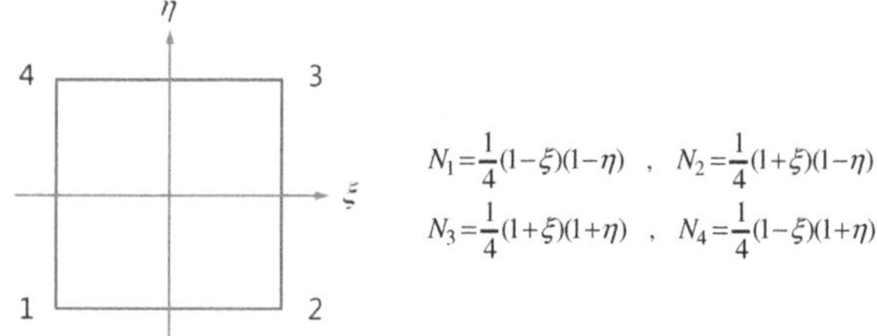

Fig. 1 Two-dimensional element and shape functions

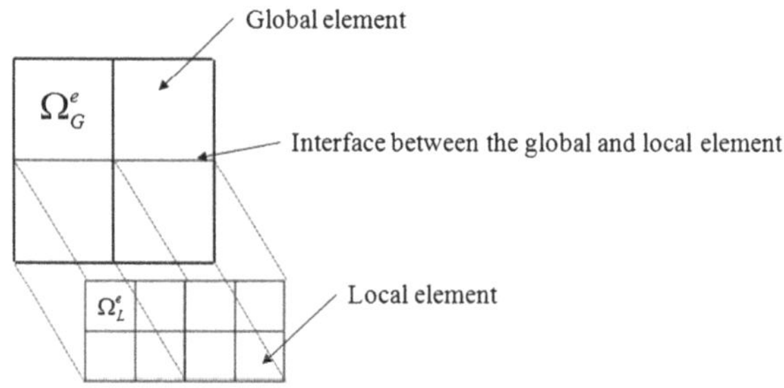

Fig. 2 Global mesh and local mesh for the superimposed finite element

$$
\mathrm{u} = \begin{cases} \mathrm{u}_G & = N_G u_G & \text{on } \Omega_G \\ \mathrm{u}_G + 0 & = N_G u_G & \text{on } \Gamma_{GL} \\ \mathrm{u}_G + \mathrm{u}_L & = N_G u_G + N_L u_L & \text{on } \Omega_L \end{cases} \tag{3}
$$

where N is the shape function, G and L are the global domain and local domain, respectively. In the global domain, the displacement vector is represented by the general interpolation function called the shape function. In case of a two-dimensional element, the shape function of the global domain is like in Fig. 1. In the local domain, the displacement is represented by the sum of the global displacement vector and local displacement vector. In this domain, a hierarchical shape function must be used as the interpolation function. The hierarchical shape function is obtained by adding new shape functions to the existing ones. In Fig. 2, the local element patch is composed of four elements. For presenting the behavior of local elements, some shape functions are added in the original shape functions like in Fig. 3. This modified shape functions and added shape functions are called the hierarchical shape function. At the interface between the global domain and the local domain, the displacement vector is represented by the global displacement only for the displacement continuity.

The strain vector which is the derivative of the displacement is represented by Eq. (4).

$$
\varepsilon = \begin{cases} \varepsilon_G & = B_G u_G & \text{on } \Omega_G \\ \varepsilon_G + 0 & = B_G u_G & \text{on } \Gamma_{GL} \\ \varepsilon_G + \varepsilon_L & = B_G u_G + B_L u_L & \text{on } \Omega_L \end{cases} \tag{4}
$$

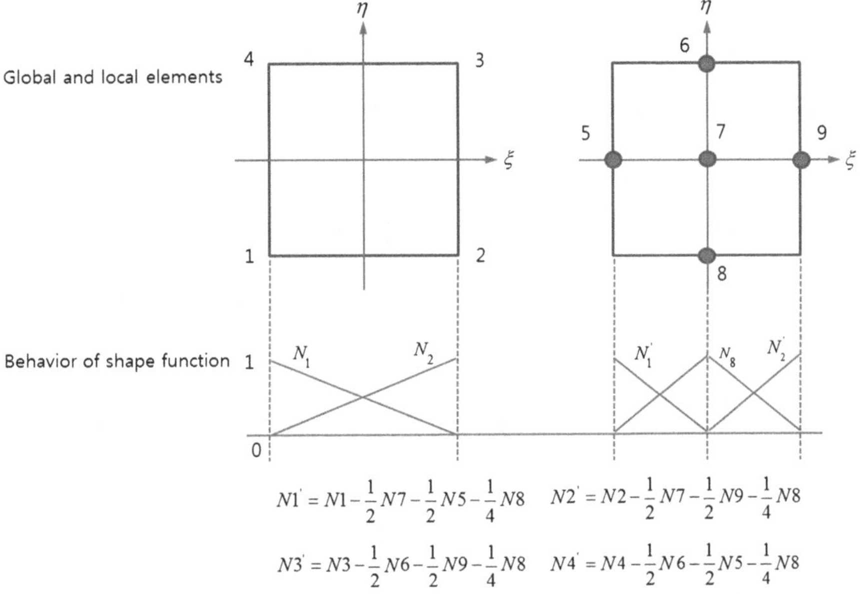

$$N1' = N1 - \frac{1}{2}N7 - \frac{1}{2}N5 - \frac{1}{4}N8 \quad N2' = N2 - \frac{1}{2}N7 - \frac{1}{2}N9 - \frac{1}{4}N8$$

$$N3' = N3 - \frac{1}{2}N6 - \frac{1}{2}N9 - \frac{1}{4}N8 \quad N4' = N4 - \frac{1}{2}N6 - \frac{1}{2}N5 - \frac{1}{4}N8$$

Fig. 3 Local element patch composed by four elements and its hierarchical shape functions in a two-dimensional element

where B_G and B_L are the derivatives of the global and local displacement interpolation functions.

For an efficient analysis, the domain is discretized as shown in Fig. 2 [17]. As the global and local element boundaries coincide, the stiffness matrix is assembled without additional numerical costs.

From Eqs. (2) and (4), the governing equations in the global domain and local domain are obtained as Eqs. (5) and (6), respectively.

$$[K_G^e]\{u_G^e\} = \{Q_G^e\}, \quad \begin{aligned} K_G^e &= \int_{\Omega_{G^e}} B_G^T \mathrm{D} B_G \mathrm{d}\Omega \\ Q_G^e &= \int_{\Gamma^e} N_G^T t \mathrm{d}\Gamma + \sum_i N_G \mathrm{f}_i \end{aligned} \tag{5}$$

$$\begin{bmatrix} K_G^e & K_C^e \\ K_C^{eT} & K_L^e \end{bmatrix} \begin{Bmatrix} u_G^e \\ u_L^e \end{Bmatrix} = \begin{Bmatrix} Q_G^e \\ Q_L^e \end{Bmatrix}, \quad \begin{aligned} K_G^e &= \int_{\Omega_L^e} B_G^T D B_G \, \mathrm{d}\Omega \\ K_C^e &= \int_{\Omega_L^e} B_G^T D B_L \, \mathrm{d}\Omega \\ K_L^e &= \int_{\Omega_L^e} B_L^T D B_L \, \mathrm{d}\Omega \\ Q_G^e &= \int_{\Gamma^e} N_G^T t \, \mathrm{d}\Gamma + \sum_i N_G \mathrm{f}_i \\ Q_L^e &= \int_{\Gamma^e} N_L^T t \, \mathrm{d}\Gamma + \sum_i N_L \mathrm{f}_i \end{aligned} \tag{6}$$

where K is the stiffness matrix and Q is the nodal load vector.

The global governing equation can be obtained from Eqs. (5) and (6). Local nodes that belong to the physical boundary are assigned the physical boundary conditions of the problem.

3 Application to the Crack Propagation Simulation

A cohesive zone model, which is one of interface modelling methods, was introduced to model the interface. The complicated behavior of the interface is characterized into one cohesive layer. The material behavior of the interface is determined by a traction-separation law. Therefore, this is easy to implement and estimates accurately the experimental results. But this has some drawbacks. First, re-meshing process is required because the cohesive elements are generated during the analysis. Second, it is difficult to generate the cohesive element because the thickness of the cohesive element is very thin. The proposed method overcomes these drawbacks of the cohesive zone model by using the superimposed finite element method. When crack propagation occurs, the crack propagated region is refined by the superimposed finite element method using the local fine mesh, as shown in Fig. 4. The local fine mesh includes the cohesive element on the crack path. In Fig. 4, the red line is the cohesive element. This local mesh is generated automatically by the predetermined dividing patterns as shown in Fig. 5. The predetermined local mesh patterns lead to easy implementation and generation of the local mesh, so it improves the efficiency of the meshing and modeling scheme.

The local mesh is generated finer than the global mesh through the hierarchical concept, as shown in Fig. 6. If the hierarchical level is increased, the local mesh in the current level is generated as a new one by dividing the local mesh in the previous level. Thus, the element density in the local region is controlled through the hierarchical level in a simple way. From this concept, the elements in the crack propagated region can be discretized sufficiently without an additional mesh generation process.

The maximum principal stress is used as the criterion of crack propagation and the bilinear model as the traction separation law is used for the cohesive zone model, as shown in Fig. 7. In this model, the total area of the bilinear model represents the fracture toughness of the interface. As the load is applied, the cohesive layer deforms linearly, as shown in Fig. 7a. After the stress in the interface reaches the critical stress, the stress decreases and the separation increases, as shown in Fig. 7b. Finally, complete failure occurs when the separation reaches δ_f.

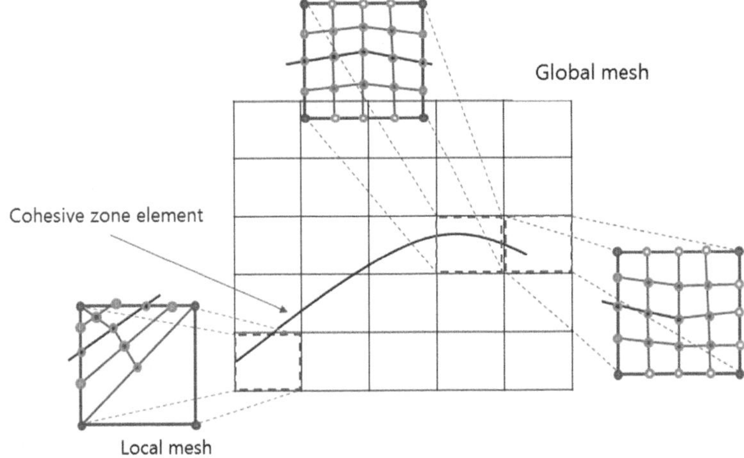

Fig. 4 Local refinement when crack propagation occured

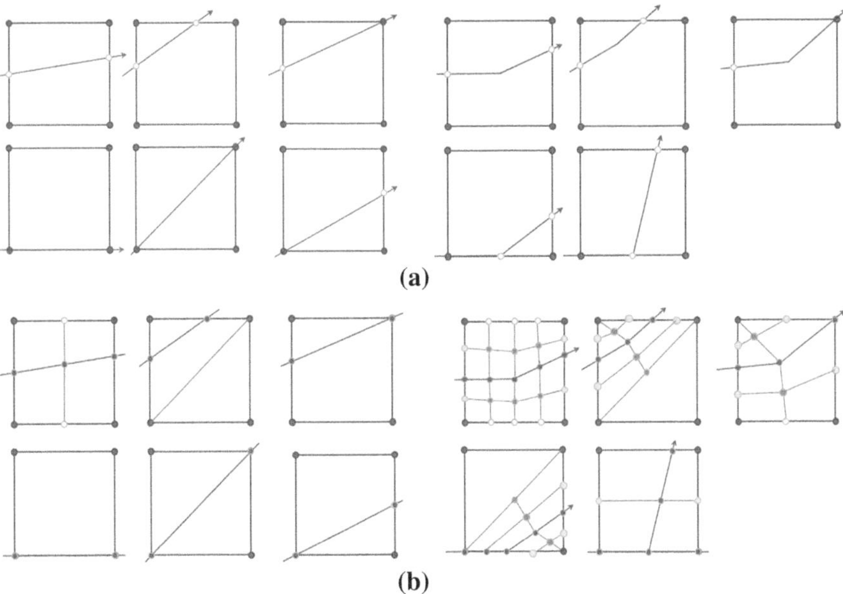

Fig. 5 The divided patterns (**a**) and the local mesh refinement by the predetermined dividing patterns (**b**)

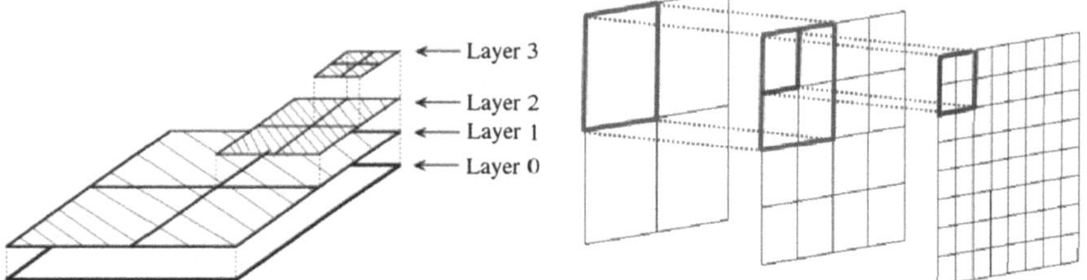

Fig. 6 The hierarchical concepts of the local mesh patch generation

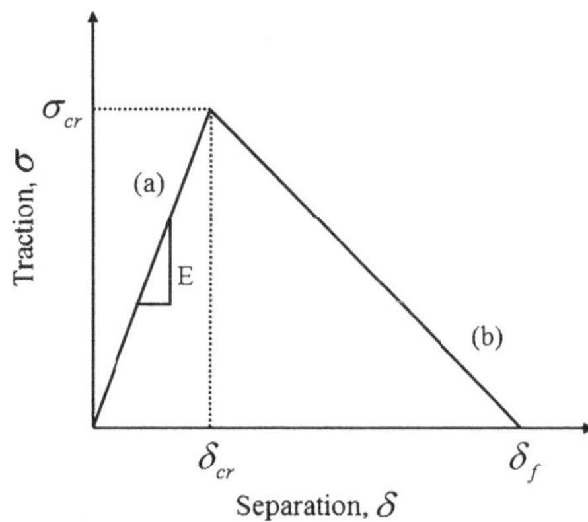

Fig. 7 The bilinear traction-separation model for the cohesive zone model

4 Numerical Examples

4.1 A Double Cantilever Beam Problem

For verification, the proposed method is applied to the double cantilever beam problem, as shown in Fig. 8. This material behavior is taken as isotropic elastic for this problem and the mode-I separation exists only in this problem. The material properties and critical energy release rate are assigned as follows:

$$E = 100\,\mathrm{N/mm^2},\ \upsilon = 0.3,\ f_u = 1.0\,\mathrm{N/mm^2},\ G_F = 0.1\,\mathrm{Nmm^{-1}} \tag{7}$$

where E is the elastic modulus, υ is the Poisson's ratio, f_u is the cohesive strength and G_F is the fracture toughness.

A finite element analysis was performed using the in-house code and the results are compared with the results given in Ref. [16], which was used the continuous partition of unity method (PUM). These results are the load-displacement curve at the loading point and the crack path is the straight line, which is perpendicular to the direction of the loading. The model is shown in Fig. 9 and the analysis results are shown in Fig. 10. The numerical results by the proposed method describe well the results of the reference.

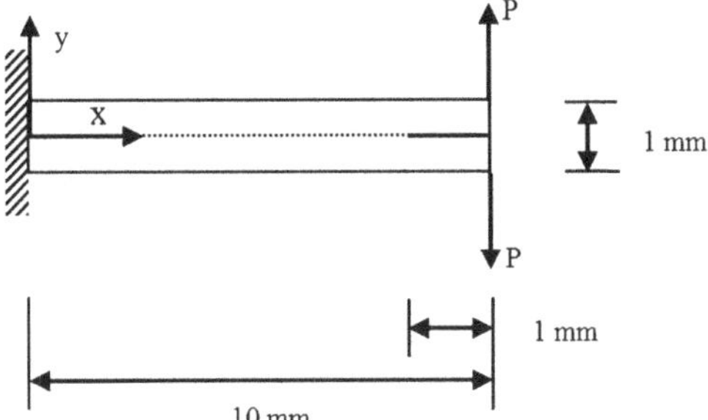

Fig. 8 The geometry of a double cantilever beam problem

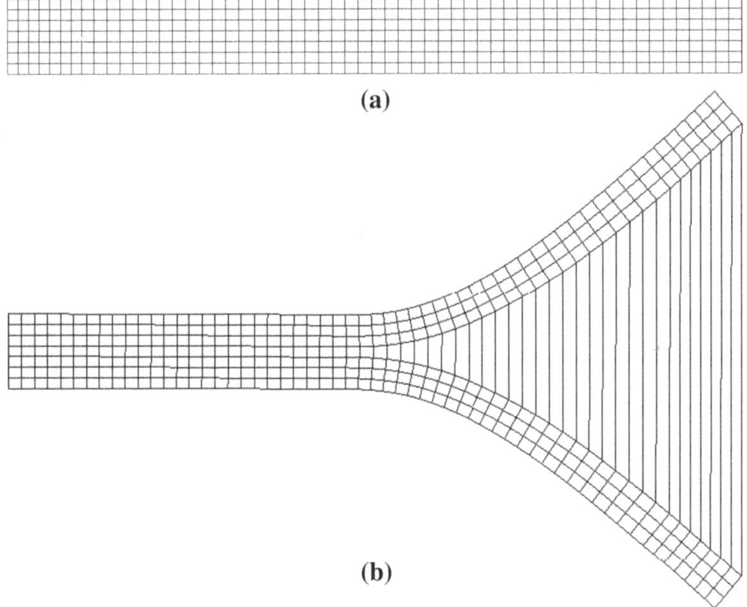

Fig. 9 The finite element model (**a**) and deformed shape (**b**) of a double cantilever beam problem

4.2 A Three Point Bending Problem

Consider the beam in three point bending shown in Fig. 11. The growth of a cohesive crack in such a specimen has been studied extensively by Carpinteri and Colombo [17] using finite element analysis and the node release technique. This load-deflection curve is shown in Fig. 12 for the following geometrical parameters (t is the specimen thickness):

$$b = 0.15\,\text{m},\ l = 4b,\ t = b,\ a = 0 \tag{8}$$

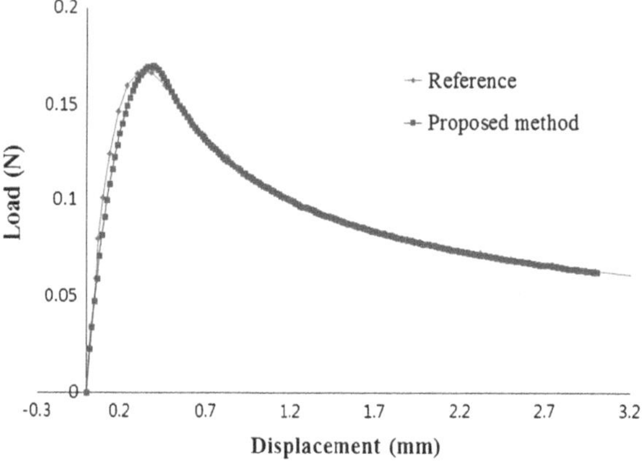

Fig. 10 The load-displacement curve as calculated using the proposed method (A double cantilever beam problem)

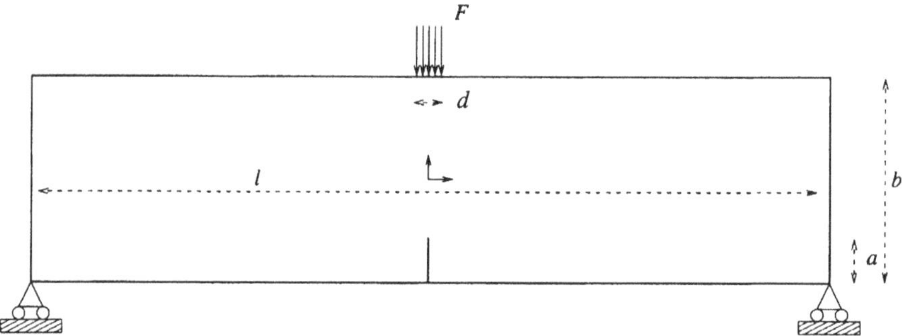

Fig. 11 The geometry of a three point bending problem

The material properties are assigned as follows

$$E = 36,500\,\text{MPa}, \upsilon = 0.1,\ f_u = 3.19\,\text{MPa},\ G_F = 50\,\text{Nm}^{-1} \tag{9}$$

where E is the elastic modulus, υ is the Poisson's ratio, f_u is the cohesive strength and G_F is the fracture toughness.

A finite element analysis was performed using the in-house code and the results are compared with the results of Carpinteri and Colombo. The model is shown in Fig. 13 and analysis results are shown in Fig. 12. The numerical results by the proposed method describe well the referenced results.

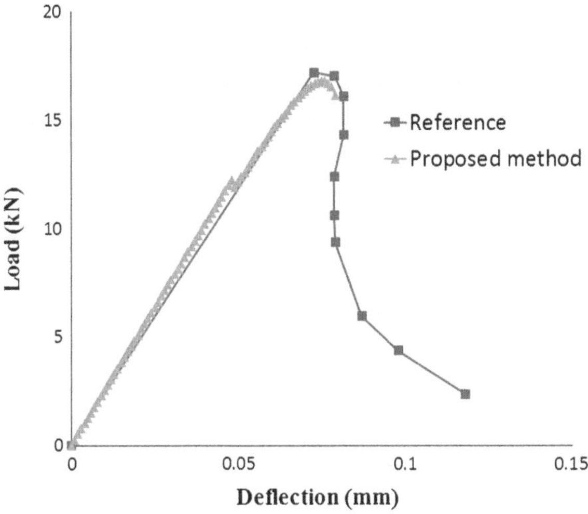

Fig. 12 The load-deflection curve at the loading point as calculated using the proposed method (A three point bending problem)

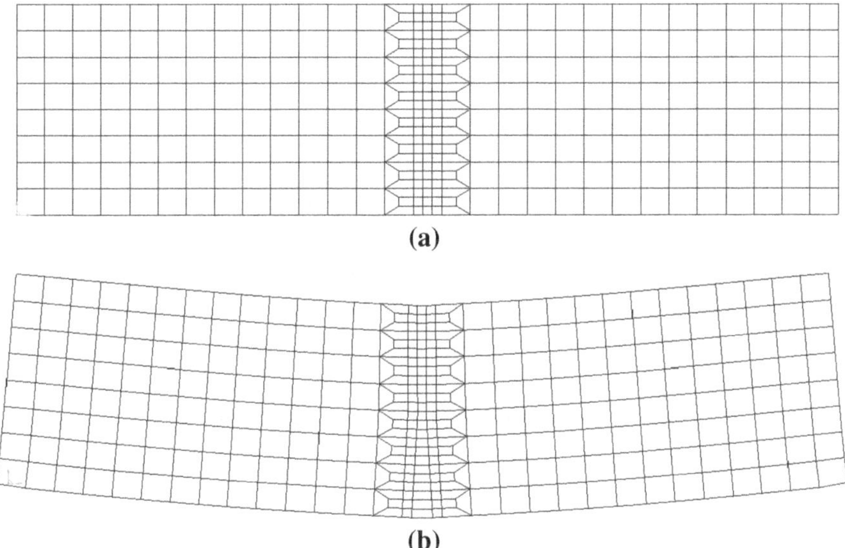

Fig. 13 The finite element model of a three point bending problem (**a**) and deformed shape (**b**)

5 Conclusion

In this study, an effective method by using the superimposed finite element method is proposed to overcome drawbacks of the cohesive zone model in the crack propagation simulation. For using the cohesive zone model in the crack propagation simulation, re-meshing process is required to insert the cohesive element. In the proposed method, the superimposed finite element method is used to overcome this drawback. When the crack propagates, the local element patch is generated using the predefined patterns and the hierarchical concept. The local element patch includes the cohesive elements on the crack surface. And then, this patch is superimposed on the global element. Therefore, the crack propagation simulation can be performed with cohesive elements efficiently.

This method is not required the re-meshing process for the crack propagation and generate the cohesive element easily. And the generation of the local mesh patch is easier and more efficient because of using the predefined patterns. Additionally, the mesh density of the local fine elements can be controlled easily due to the hierarchical concept. Consequently, the proposed method improves the efficiency of the crack propagation simulation.

The proposed method was applied to a double cantilever beam problem. The numerical results using the proposed method describe well reference results. The proposed method is also applied to a three point bending problem and the results are compared with reference results. The numerical results using the proposed method describe well the reference one.

References

1. Charalambides, P.G., Mcmeeking, R.M.: Finite element method simulation of crack propagation in a brittle microcracking solid. Mech. Mater. **6**, 71–87 (1987)
2. Apalak, M.K., Engin, A.: An investigation on the initiation and propagation of damaged zones in adhesively bonded lap joints. J. Adhes. Sci. Technol. **17**, 1889–1921 (2003)
3. Moës, N., Dolbow, J., et al.: A finite element method for crack growth without remeshing. Int. J. Numer. Methods Eng. **46**, 131–150 (1999)
4. Moës, N., Belytschko, T.: Extended finite element method for cohesive crack growth. Eng. Fract. Mech. **69**, 813–833 (2002)
5. Zi, G., Belytschko, T.: New crack-tip elements for XFEM and applications to cohesive cracks. Int. J. Numer. Meth. Eng. **57**, 2221–2240 (2003)
6. Unger, J.F., Eckardt, S., et al.: Modelling of cohesive crack growth in concrete structures with the extended finite element method. Comput. Methods Appl. Mech. Eng. **196**, 4087–4100 (2007)
7. Elices, M., Guinea, G.V., et al.: The cohesive zone model : advantages, limitations and challenges. Eng. Fract. Mech. **69**, 137–163 (2001)
8. Alfano, M., Furgiuele, A., et al.: Mode I fracture of adhesive joints using tailored cohesive zone models. Int. J. Fract. **157**, 193–204 (2009)
9. Fish, J.: The s-version of the finite element method. Comput. Struct. **43**, 539–547 (1992)
10. Fish, J., Guttal, R.: The s-version of finite element method for laminated composites. Int. J. Numer. Methods Eng. **39**, 3341–3662 (1996)
11. Yue, Z., Robbins, D.H.: Adaptive superposition of finite element meshes in elastodynamic problems. Int. J. Numer. Methods Eng. **63**, 1604–1635 (2005)
12. Lee, H.S., Song, J.H., et al.: Combined extended and superimposed finite element method for cracks. Int. J. Numer. Methods Eng. **59**, 1119–1136 (2004)
13. Wang, S., Wang, M.Y.: A moving superimposed finite element method for structural topology optimization. Int. J. Numer. Methods Eng. **65**, 1892–1922 (2006)

14. Kim, Y.T., Lee, M.J., et al.: Simulation of adhesive joints using the superimposed finite element method and a cohesive zone model. Int. J. Adhes. Adhes. **31**, 357–362 (2011)
15. Park, J.W., Hwang, J.W., et al.: Efficitne finite element analysis using mesh superposition technique. Finite Elem. Anal. Des. **39**, 619–638 (2003)
16. Remmers, J.J.C., de Borst, R., et al.: A cohesive segments method for the simulation of crack growth. Comput. Mech. **31**, 69–77 (2003)
17. Carpinteri, A., Colombo, G.: Mumerical analysis of catastrophic softening behaviour (snap-back instability). Comput. Struct. **31**, 607–636 (1989)

A Parametric Finite-Volume Formulation for Linear Viscoelasticity

Severino P. C. Marques, Romildo S. Escarpini Filho and Guillermo J. Creus

Abstract This chapter of contribution presents a new numerical model for the analysis of structures of heterogeneous materials with linear viscoelastic constituents. The model is based on the recently developed parametric finite-volume theory. The use of quadrilateral subvolumes made possible by the mapping facilitates efficient modeling of microstructures with arbitrarily shaped heterogeneities, and eliminates artificial stress concentrations produced by the rectangular subvolumes employed in the standard version. The parametric formulation is here extended to model viscoelastic behavior. Several examples, including both homogeneous and heterogeneous situations, are analyzed. Comparison between numerical and analytical results has shown an excellent performance of the proposed model.

1 Introduction

The application of heterogeneous materials to fill the needs of diverse industrial sectors has significantly increased in the last years [1]. Some of these composites are constituted by high performance fibers (carbon, glass, metal, ceramic, etc.) embedded in a polymeric matrix. Due to the time-dependent behavior of polymers,

S. P. C. Marques (✉) · R. S. Escarpini Filho
LCCV/Center of Technology Federal University of Alagoas,
Campus AC Simoes, Maceio-AL, Brazil
e-mail: marques.severino@gmail.com

R. S. Escarpini Filho
e-mail: romildo@lccv.ufal.br

G. J. Creus
CEMACOM-UFRGS, Porto Alegre-RS, Brazil
e-mail: creus@ufrgs.br

A. Öchsner et al. (eds.), *Design and Analysis of Materials and Engineering Structures*,
Advanced Structured Materials 32, DOI: 10.1007/978-3-642-32295-2_6,

the composites exhibit an effective viscoelastic response that is affected by environmental agents, particularly temperature, humidity and age [2–4].

On the other hand, composites show a heterogeneous texture on the microlevel, determined by the constitutive behavior of the matrix material and the embedded fibers.

Approximate constitutive relations may be obtained using micromechanical models [5, 6]. Traditionally, the finite element method is the numerical technique most employed to analyse composites, particularly to understand their macro-level behavior [7, 8]. To have a more complete understanding of the viscoelastic behavior of a composite, the influence of the heterogeneous nature at the microscale on the macroscale response must be examined in detail. Lately, computational tools other than finite elements have been proposed to model heterogeneities and their effects under a variety of thermo-mechanical conditions [9–11]. An attractive alternative technique is the recently developed parametric formulation of the finite-volume theory [12], which incorporates a parametric mapping capability into the Finite-Volume Direct Averaging Method (FVDAM) [11]. The use of quadrilateral subvolumes made possible by the parametric mapping facilitates the modeling of heterogeneities and eliminates artificial stress concentrations produced by the rectangular subvolumes employed in the standard version. Thus, the procedure offers flexibility and accuracy comparable to the finite element method for modeling of structures made of material with microstructure constituted by arbitrarily shaped heterogeneities [13, 14].

This parametric formulation of the finite-volume theory, initially formulated for thermo-mechanical analysis of linear elastic composite materials [12], has been extended to include plastic effects and used as a framework for the development of a homogenization technique applied to materials with periodic heterogeneous microstructures [15].

In the present chapter a numerical model for the analysis of heterogeneous materials with linear viscoelastic constituents is presented. The model has been derived on the parametric finite-volume theory framework and employs a state variables formulation for the computation of the viscoelastic strains [16].

Several examples, including both homogenous and heterogeneous materials, were analyzed and their results were compared with analytical solutions found in the literature or obtained by using the correspondence principle. The comparison between the numerical and analytical results has shown an excellent performance of the proposed model.

2 Parametric Viscoelastic Formulation of the Finite-Volume Theory: Plane Stress and Plane Strain Problems

In the parametric finite-volume theory, the whole domain occupied by the heterogeneous material is discretized into quadrilateral subvolumes [12]. The formulation is based on the mapping of a reference square subvolume onto each quadrilateral subvolume, as shown in Fig. 1.

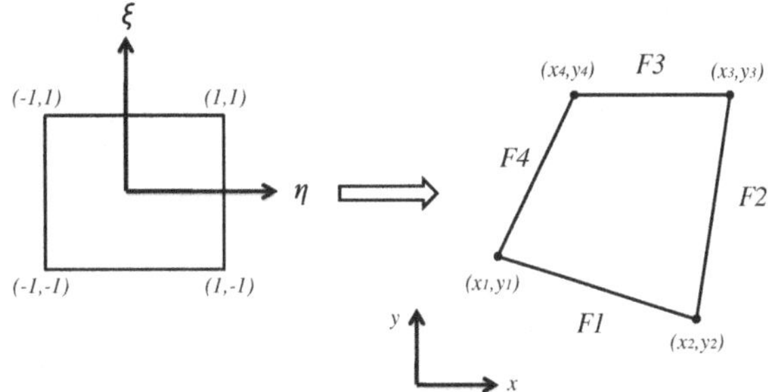

Fig. 1 Mapping of the reference square subvolume in the $\eta - \xi$ plane onto a quadrilateral subvolume in the x–y plane of the actual microstructure

The displacement field at time t is approximated using a second order expansion in the reference square subvolume coordinates as follows

$$\begin{aligned} u_1(t) &= U^t_{1(00)} + \eta U^t_{1(10)} + \xi U^t_{1(01)} + \frac{1}{2}\left(3\eta^2 - 1\right)U^t_{1(20)} + \frac{1}{2}\left(3\xi^2 - 1\right)U^t_{1(02)} \\ u_2(t) &= U^t_{2(00)} + \eta U^t_{2(10)} + \xi U^t_{2(01)} + \frac{1}{2}\left(3\eta^2 - 1\right)U^t_{2(20)} + \frac{1}{2}\left(3\xi^2 - 1\right)U^t_{2(02)} \end{aligned} \tag{1}$$

where $U^t_{i(kl)}$ are the polynomial coefficients.

The formulation in the viscoelastic procedure is similar to that in the elastic case, which may be seen in [12]. For an isotropic linear viscoelastic material occupying the subvolume, the increments of stress are related to increments of strain, corresponding to time interval $[t, t + \Delta t]$, through the expression

$$\begin{Bmatrix} \Delta\sigma_{11} \\ \Delta\sigma_{22} \\ \Delta\sigma_{12} \end{Bmatrix} = \overline{\boldsymbol{C}}\left(\begin{Bmatrix} \Delta\varepsilon_{11} \\ \Delta\varepsilon_{22} \\ \Delta\varepsilon_{12} \end{Bmatrix} - \begin{Bmatrix} \Delta\varepsilon^T_{11} \\ \Delta\varepsilon^T_{22} \\ 0 \end{Bmatrix} - \begin{Bmatrix} \Delta\varepsilon^V_{11} \\ \Delta\varepsilon^V_{22} \\ \Delta\varepsilon^V_{12} \end{Bmatrix}\right) \tag{2}$$

where the indices T and V are used to indicate the increments of thermal and viscoelastic strains respectively and $\overline{\boldsymbol{C}}$ is the linear elastic constitutive matrix for plane stress and plane strain. The surface-averaged incremental stress–strain relation for the subvolume faces can be expressed in the form

$$\begin{aligned} \left.\begin{Bmatrix} \langle\Delta\sigma_{11} \\ \langle\Delta\sigma_{22} \\ \langle\Delta\sigma_{12} \end{Bmatrix}\right|_{k=2,4} &= \overline{\boldsymbol{C}}\left.\left(\begin{Bmatrix} \langle\Delta\varepsilon_{11} \\ \langle\Delta\varepsilon_{22} \\ \langle\Delta\varepsilon_{12} \end{Bmatrix} - \begin{Bmatrix} \langle\Delta\varepsilon^T_{11} \\ \langle\Delta\varepsilon^T_{22} \\ 0 \end{Bmatrix} - \begin{Bmatrix} \langle\Delta\varepsilon^V_{11} \\ \langle\Delta\varepsilon^V_{22} \\ \langle\Delta\varepsilon^V_{12} \end{Bmatrix}\right)\right|_{k=2,4} \\ \left.\begin{Bmatrix} \langle\Delta\sigma_{11} \\ \langle\Delta\sigma_{22} \\ \langle\Delta\sigma_{12} \end{Bmatrix}\right|_{k=1,3} &= \overline{\boldsymbol{C}}\left.\left(\begin{Bmatrix} \langle\Delta\varepsilon_{11} \\ \langle\Delta\varepsilon_{22} \\ \langle\Delta\varepsilon_{12} \end{Bmatrix} - \begin{Bmatrix} \langle\Delta\varepsilon^T_{11} \\ \langle\Delta\varepsilon^T_{22} \\ 0 \end{Bmatrix} - \begin{Bmatrix} \langle\Delta\varepsilon^V_{11} \\ \langle\Delta\varepsilon^V_{22} \\ \langle\Delta\varepsilon^V_{12} \end{Bmatrix}\right)\right|_{k=1,3} \end{aligned} \tag{3}$$

Using the Cauchy equation, the surface-averaged increments of tractions on the faces of the subvolume are given as

$$\left\{ \begin{matrix} \Delta t_1 \\ \Delta t_2 \end{matrix} \right\}\bigg|_{k=1,2,3,4} = \left(\begin{bmatrix} n_1 & 0 & n_2 \\ 0 & n_2 & n_1 \end{bmatrix} \left\{ \begin{matrix} \langle \Delta\sigma_{11} \\ \langle \Delta\sigma_{22} \\ \langle \Delta\sigma_{12} \end{matrix} \right\} \right)\Bigg|_{k=1,2,3,4} \tag{4}$$

n_1 and n_2 being the components of the unit normal vector to each face F_k in the directions x and y, respectively.

The global incremental stiffness matrix K_G is obtained using the same procedures employed in [12]. The local stiffness matrices are assembled into a global system of equations by applying continuity conditions for the increments of surface-averaged interfacial tractions and displacements, followed by the specified boundary conditions. The system of equations, resulting from this approach, relates the increments of unknown interfacial and boundary surface-averaged displacements to the increments of surface-averaged tractions, as follows

$$\boldsymbol{K}_G \Delta \overline{\boldsymbol{U}} = \Delta \bar{\boldsymbol{t}} + \Delta \bar{\boldsymbol{t}}_{\boldsymbol{o}} \tag{5}$$

In Eq. (5), the vector $\Delta\overline{\boldsymbol{U}}$ contains the increments of unknown interfacial and boundary surface-averaged displacements, $\Delta\bar{\boldsymbol{t}}$ is the vector of the net increments of surface-averaged tractions along the internal interfaces and the discretized boundary and $\Delta\bar{\boldsymbol{t}}_{\boldsymbol{o}}$ indicates a global incremental loading vector with information on body forces and thermal and viscoelastic effects. The vectors $\Delta\bar{\boldsymbol{t}}$ and $\Delta\bar{\boldsymbol{t}}_{\boldsymbol{o}}$ are known at the beginning of each time incremental step. Due to the interfacial traction continuity conditions on the common interfaces of adjacent subvolumes, $\Delta\bar{\boldsymbol{t}}$ is sparse with its nonzero terms being the increments of surface-averaged tractions on the discretized boundary regions.

3 Evaluation of the Viscoelastic Strains

3.1 Linear Viscoelastic Constitutive Relation

The time-dependent strain components are determined incrementally using the state variable approach [3, 16] that avoids the need to carry on the whole past history of the deformation.

For an isotropic material and considering the volumetric and distortional effects separately, the constitutive viscoelastic relations can be written in the form

$$\begin{aligned} \varepsilon_o(t) &= D_K(0)\sigma_m(t) - \int_0^t \frac{\partial D_K(t-\tau)}{\partial \tau}\sigma_m(\tau)d\tau \\ e_i(t) &= D_G(0)s_i(t) - \int_0^t \frac{\partial D_G(t-\tau)}{\partial \tau}s_i(\tau)d\tau \end{aligned} \tag{6}$$

where ε_o, e_i, σ_m and s_i stand for the volumetric strain, deviator strain components, hydrostatic stress and deviator stress components, respectively. D_K and D_G are the material creep functions corresponding to the volumetric and distortional behaviors, respectively. $D_K(0) = 1/K$ and $D_G(0) = 1/2G$ being K and G the elastic values of bulk and shear moduli, respectively.

The creep functions in (6) can be approximated by a Dirichlet-Prony series as

$$\begin{aligned} D_K(t-\tau) &= D_K^0 + \sum_{p=1}^{n} D_K^p \left[1 - \mathrm{e}^{-\left(\frac{t-\tau}{\theta_K^p}\right)}\right] \\ D_G(t-\tau) &= D_G^0 + \sum_{p=1}^{n} D_G^p \left[1 - \mathrm{e}^{-\left(\frac{t-\tau}{\theta_G^p}\right)}\right] \end{aligned} \tag{7}$$

where D_K^0, D_K^p, D_G^0, D_G^p, θ_K^p *and* θ_G^p are material constants. Through the same reasoning used above, the elastic and viscoelastic parts of the volumetric and deviator strains can be given by

$$\begin{aligned} \varepsilon_o^e(t) &= D_K(0)\sigma_m(t) \quad \varepsilon_o^V(t) = \sum_{p=1}^{n} \varphi_K^p(t) \\ e_i^e(t) &= D_G(0)s_i(t) \quad e_i^V(t) = \sum_{p=1}^{n} \varphi_{Gi}^p(t) \end{aligned} \tag{8}$$

In Eq. (8), φ_K^p and φ_{Gi}^p indicate the state variables corresponding to the volumetric and distortional behaviors which are defined by

$$\varphi_K^p(t) = \frac{D_K^p}{\theta_K^p}\int_0^t e^{-\left(\frac{t-\tau}{\theta_K^p}\right)}\sigma_m(\tau)d\tau \quad \varphi_{Gi}^p(t) = \frac{D_G^p}{\theta_G^p}\int_0^t e^{-\left(\frac{t-\tau}{\theta_G^p}\right)}s_i(\tau)d\tau \tag{9}$$

Considering a small time interval Δt and assuming $\sigma_j(\tau)$ as constant along the interval Δt and equal to $\sigma_j(t)$ the expression that allows obtaining the variables φ_{ij}^p at time $t + \Delta t$ as functions of their values at time t can be easily derived from Eq. (9)

$$\varphi_{ij}^p(\mathrm{t}+\Delta\mathrm{t}) = \varphi_{ij}^p(\mathrm{t})\mathrm{e}^{-\left(\frac{\Delta\mathrm{t}}{\theta_{ij}^p}\right)} + D_{ij}^p\left[1 - \mathrm{e}^{-\left(\frac{\Delta\mathrm{t}}{\theta_{ij}^p}\right)}\right]\sigma_j(\mathrm{t}) \tag{10}$$

This simple procedure is adequate as long as the material relaxation times are not too short.

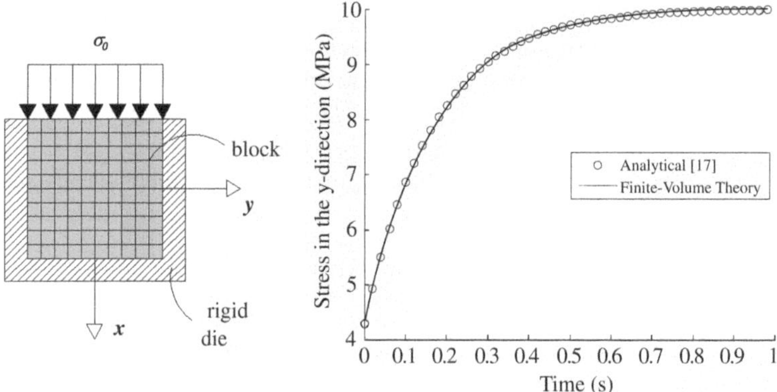

Fig. 2 Viscoelastic block in a rigid die. Plot of the horizontal compressive stress versus time

4 Numerical Examples

4.1 Viscoelastic Block in a Rigid Die

A viscoelastic block confined inside an infinitely rigid container is subjected to an instantaneously applied and constant vertical pressure $\sigma_0 = 10$ MPa as showed in Fig. 2 [17]. The block is made of an isotropic and homogeneous material with elastic behavior in dilatation and viscoelastic behavior in shear. The friction effect between the block and the container is neglected. The value of the normal stress in the y-direction is the unknown. Figure 2 shows also the discretization used.

The viscoelastic behavior was modeled by a Maxwell element with a spring of constant $G = 3846.154$ MPa and a dashpot with $\eta_G = 400$ MPa.s. The adopted elastic Bulk modulus was $K = 8333.333$ MPa. As the formulation employs a generalized Kelvin model, the Maxwell model was simulated by a standard model with a very small spring constant for the Kelvin element.

The analytical solution for this example is given by [17]

$$\sigma_y(t) = -\sigma_o \left\{ 1 - \frac{6G}{3K + 4G} e^{-\left[\frac{3K}{(3K+4G)\theta}\right] t} \right\} \tag{11}$$

where $\theta = \eta_G/G$. Figure 2 presents both the numerical and analytical results for the horizontal compression stress as function of time t. It is observed that the proposed formulation provides results nearly identical with the exact analytical solution. The absolute maximum errors were about 1.75 % for a time increment $\Delta t = 0.02$ s and 0.08 % for $\Delta t = 0.001$ s.

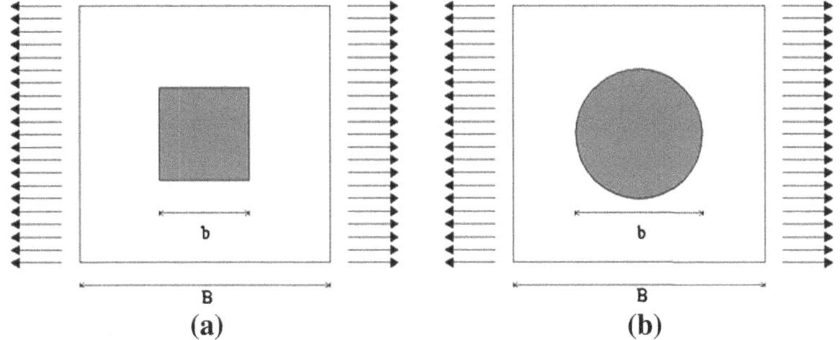

Fig. 3 Composite with different fiber shapes: **a** *square* and **b** *circular*

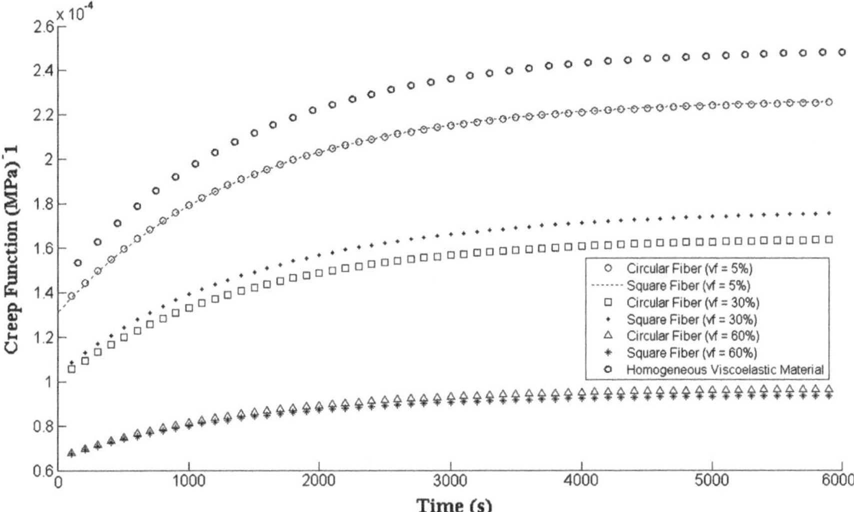

Fig. 4 Effective creep functions for different volume fractions and fiber shapes

4.2 Elastic Inclusion Embedded in a Viscoelastic Matrix

Figure 3 shows a problem which consists of an elastic inclusion embedded in a viscoelastic matrix subjected to a uniform tensile loading given by $\sigma_0 = 100$ MPa The composite was assumed in plane stress state and with total dimensions of 100×100 mm. Due to the symmetry of the problem, only one quarter of the domain was modeled.

To verify the influence of the elastic inclusion on the viscoelastic response of the composite, circular and square geometrical shapes with three different volume fractions were considered. The matrix was modeled using a standard model with the following mechanical properties: $E_1 = 7000$ MPa, $\nu = 0.3$, $E_2 = 9333.3$ MPa and $\eta_2 = 1.3611 \times 10^7$ $MPa.s$. The elastic properties of the inclusion were $E = 30250$ MPa and $v = 0.443$ Volume fractions v_f of 5, 30 and 60 % were adopted for each inclusion shape. The effective creep functions obtained in the analyses are showed in Fig. 4. As expected, the presence of the elastic inclusion reduces the effective viscoelastic behavior which decreases with the increase of the inclusion volume fraction. It is also observed in Fig. 4 that the influence of the inclusion shape on the composite effective behavior depends on the inclusion volume fraction. For the smallest volume fraction, $v_f = 5$ %, there is no significant difference between the effective creep functions corresponding to the two inclusion shapes, whereas for the intermediate value, $v_f = 30$ %, it is relatively strong. Considering the creep curves for the three volume fractions, it is possible to conclude that there is a value of v_f between 5 and 60 % for which that difference is maximized.

5 Conclusion

A new numerical incremental model for the analysis of composite structures made of linear viscoelastic constituents has been developed, using the framework of the recent parametric formulation of the finite-volume theory. The model demonstrated computational efficiency and a potential to describe the behavior of different cases of linear viscoelastic structures. Several examples were analyzed and comparisons of results with analytical solutions confirm the excellent performance of the proposed model. On the other hand, considering the recognized capability of the parametric formulation of the finite-volume theory in modeling of heterogeneous solids, it is believed that the numerical tool developed can be employed to predict the viscoelastic behavior of complex multiphase composite structures.

Acknowledgments The authors would like to gratefully acknowledge the support of the Brazilian federal agencies CNPq and CAPES.

References

1. Chung, D.D.L.: Composite materials: functional materials for modern technologies. Springer, London (2004)
2. Flaggs, D.L., Crossman, F.W.: Analysis of the viscoelastic response of composite laminates during hygrothermal exposure. J. Compos. Mater. **15**, 21–40 (1981)
3. Marques, S.P.C., Creus, G.J.: Geometrically nonlinear finite elements analysis of viscoelastic composite materials under mechanical and hygrothermal loads. Comput. Struct. **53**(2), 449–456 (1994)
4. Christensen, R.M.: Viscoelastic properties of heterogeneous media. J. Mech. Phys. Solids **7**, 23–41 (1969)
5. Nemat-Nasser, S., Hori, M.: Micromechanics: Overall Properties of Heterogeneous Materials. Elsevier, Amsterdam (1999)

6. Yeong-Moo, Y., Sang-Hoon, P., Sung-Kie, Y.: Asymptotic homogenization of viscoelastic composites with periodic microstructures. Int. J. Solids Struct. **35**, 2039–2055 (1998)
7. Brinson, L.C., Knauss, W.G.: Finite element analysis of multiphase viscoelastic solids. J. Appl. Mech. **59**(4), 730–737 (1992)
8. Barbero, E.J.: Finite Element Analysis of Composite Materials. CRC Press, Boca Raton (2007)
9. Payley, M., Aboudi, J.: Micromechanical analysis of composites by the generalized methods of cells. Mech. Mater. **14**(1), 127–139 (1992)
10. Aboudi, J., Pindera, M.-J., Arnold, S.M.: Higher-order theory for functionally graded materials. Compos. Part B **30**(8),772–832 (1999)
11. Bansal, Y., Pindera, M.-J.: Efficient reformulation of the thermoelastic higher-order theory for FGMs. J. Therm. Stress. **26**(11/12), 1055–1092 (2003)
12. Cavalcante, M.A.A., Marques, S.P.C., Pindera, M.-J.: Parametric formulation of the finite-volume theory for functionally graded materials-Part I: analysis. J. Appl. Mech. **74**(5), 935–945 (2007)
13. Cavalcante, M.A.A., Marques, S.P.C., Pindera, M.-J.: Parametric formulation of the finite-volume theory for functionally graded materials-Part II: numerical results. J. Appl. Mech. **74**(5), 946–957 (2007)
14. Cavalcante, M.A.A., Marques, S.P.C., Pindera, M.-J.: Computational aspects of the parametric finite-volume theory for functionally graded materials. Comput. Mater. Sci. **44**, 422–438 (2008)
15. Khatam, H., Pindera, M.-J.: Parametric finite-volume micromechanics of periodic materials with elastoplastic phases. Int. J. Plast. **25**(7), 1386–1411 (2009)
16. Creus, G.J.: Lecture Notes in Engineering: Viscoelasticity—Basic Theory and Applications to Concrete Structures. Springer, Berlin (1986)
17. Flugge, W.: Viscoelasticity. Springer, New York (1975)

Methodology for the Quantitative Evaluation of the Structure in Cast Magnesium Alloys

Tomasz Rzychoń

Abstract Magnesium alloys with alkaline earth elements are attractive materials for components working at elevated temperature (ca. 180 °C) in the automotive industry. It is well known that properties of engineering materials depends on the microstructure. Therefore, complex procedures for the quantitative description of the microstructure, which enable to obtain repeatable and unequivocal results, are very important in process control of technology parameters as well as to determine the relationship between properties and microstructure of materials. This chapter presents a comprehensive procedure consistent with modern quality assurance systems for the quantitative assessment of primary grains and intermetallic compounds in magnesium alloys containing aluminum and strontium. The presented procedure for this alloy includes: a methodology of metallographic specimens preparation, a methodology of image acquisition with light microscopy and finally automatic image analysis operations sequence enabling detection of grain boundaries and intermetallic compounds. Moreover, the procedure contains guidelines for statistical analysis which allows for an objective interpretation of results.

1 Introduction

Calcium and strontium are important elements in advanced magnesium alloys, which are relevant for weight savings in the automotive industry or other applications [4]. The applications of most common magnesium alloys, such as AZ91

T. Rzychoń (✉)
Faculty of Materials Science and Metallurgy, Silesian University of Technology,
Krasińskiego 8, 40-019 Katowice, Poland
e-mail: tomasz.rzychon@polsl.pl

A. Öchsner et al. (eds.), *Design and Analysis of Materials and Engineering Structures*, Advanced Structured Materials 32, DOI: 10.1007/978-3-642-32295-2_7,

(Mg–9Al–0.8Zn) and AM50 (Mg–5Al–0.5Mn) with outstanding mechanical properties and die castability are limited to temperatures below 120 °C. This limitation is attributed to the low hardness of the intermetallic phase $Mg_{17}Al_{12}$ under high temperature [8]. The search for creep-resistant alloys has led to the development of rare earth containing magnesium alloys, for example AE42 (Mg–4Al–2RE, AE44 (Mg–4Al–4RE) and magnesium alloys with silicon—AS21 (Mg–2Al–1Si), strontium—AJ62 (Mg–6Al–2Sr) or calcium—AX51 (Mg–5Al–1Ca). In these alloys, the structure is characterized by second phases from the systems Al–Si, Al–RE, Al–Sr or Al–Ca at the grain boundaries which are stable compounds with a relatively high melting point [4]. The new series of Mg–Al–Ca–Sr alloys offer excellent creep resistance up to 180 °C and low cost compared to magnesium alloys containing yttrium or rare earth metals, which are conventionally used to improve the heat resistance [9]. Typical alloys representative of the Mg–Al–Ca–Sr system are the commercial MRI-153 and MRI-230D alloys, which are designated for die casting technology. Both alloys were developed by Dead Sea Magnesium and Volkswagen AG for high temperature applications, such as engine blocks and automatic transmission cases, where the operating temperatures can be as high as 175 and 250 °C, respectively. These alloys contain aluminum to ensure their yield strength and castability, while calcium and strontium are added to form intermetallic phases at grain boundaries and in the grain interior to improve creep resistance [1, 2, 5, 7,10]. In the present study, a quantitative procedure for the determination of phase content and grain size in cast MRI-230D magnesium alloy was developed.

2 Experimental

Ingots of MRI-230D alloy was the material for the research. This alloy were purchased from Dead Sea Magnesium Ltd. The chemical composition of this alloy is provided in Table 1. The content of iron, nickel and cooper is below 0.001 wt%.

Metallographic specimens were prepared according to the procedure recommended for magnesium-base alloys by the Buehler expert system [3]. Specimens for microstructure were taken from the central part of a cast billet (75 mm diameter) and etched with 3 % nitric acid in ethanol. The microstructure was characterized by optical microscopy (Olympus GX-70) and a scanning electron microscopy (Hitachi S3400) equipped with an X-radiation detector EDS (VOYAGER of NORAN INSTRUMENTS). EDS analyses were performed with an accelerating voltage of 25 keV. The phase structure of the investigated alloy was identified by X-ray diffraction (JDX-75) using Cu Kα radiation. For a quantitative description of the structure, stereological parameters describing the size and shape of the solid solution grains and phase precipitations were selected. To measure the stereological parameters, a program for image analysis "Met-Ilo" was used [6].

Table 1 Chemical composition of the MRI-230D alloy in wt%

Al	Zn	Mn	Ca	Sr	Sn	Mg
6.8	0.01	0.23	1.91	0.25	0.5	Balance

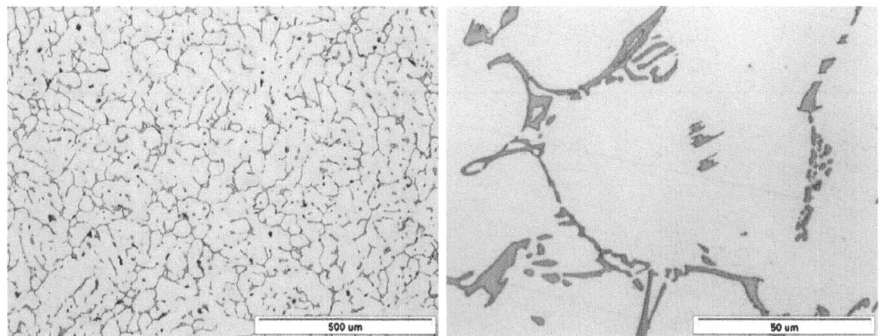

Fig. 1 Optical micrographs of MRI-230D alloy (without etching)

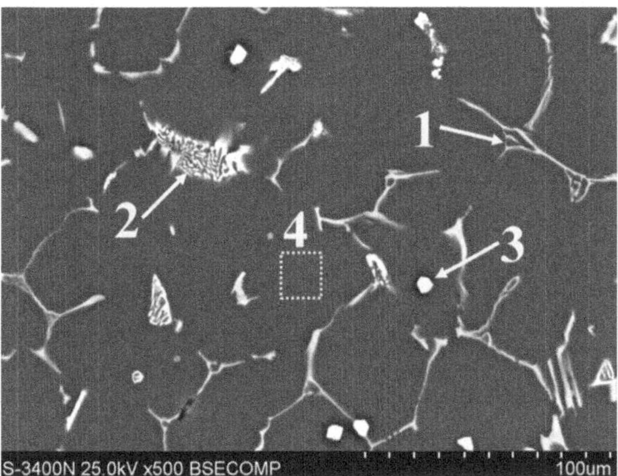

Fig. 2 Backscattered electrons (BSE) image of MRI-230D alloy

3 Results and Discussion

3.1 As Cast Microstructure

The as-cast microstructure of the MRI-230D alloy consists of a dendritic α-Mg matrix and interdendritic second-phases distributed at grain boundaries (Fig. 1).

SEM observations linked with EDS microanalysis revealed the presence of three sorts of phases in the interdendritic regions (Fig. 2). The first one was a coarse irregular-shaped eutectic. The results of EDS microanalysis showed that

Table 2 Average compositions of the intermetallic compounds and solid solution in the MRI-230D alloy analyzed by EDS (SEM)—from Fig. 2

Point	Composition (at.%)				
	Mg	Al	Ca	Mn	Sr
1	28.3	52.1	18.8	–	0.8
2	41.6	47.0	1.7	–	9.8
3	9.0	49.6	–	41.4	–
4	96.3	3.7	–	–	–

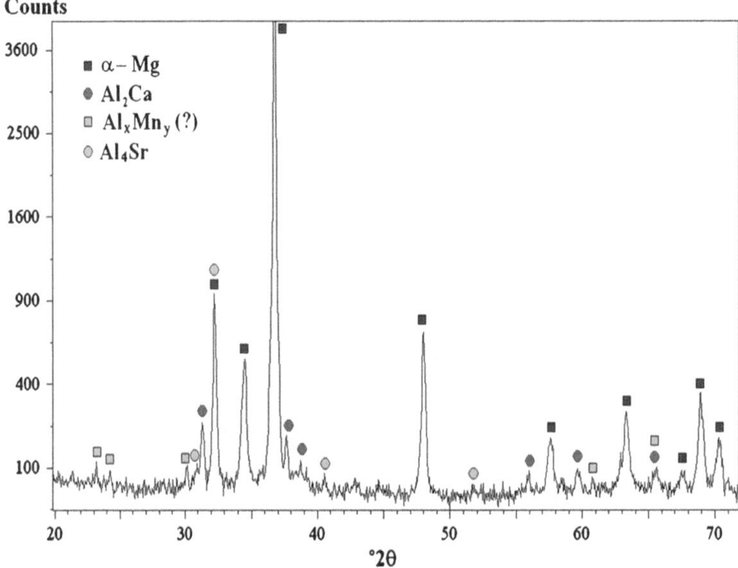

Fig. 3 XRD pattern of MRI-230D alloy

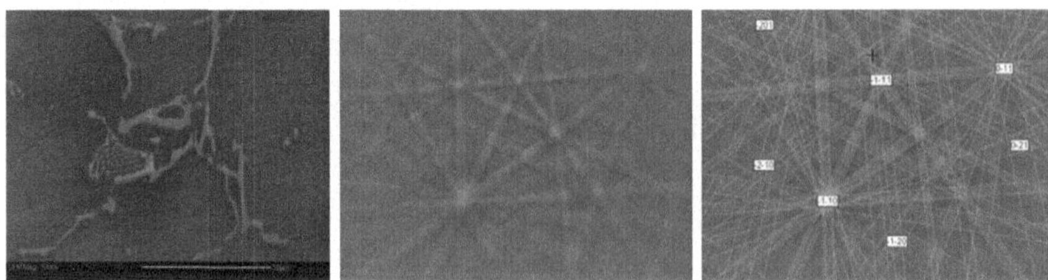

Fig. 4 Experimental EBSD pattern obtained from the irregular-shaped phase indexed as Al_2Ca (cubic crystal structure), Mean Angle of Deviation (MAD) = 0.40°, orientation = (135.7, 31.5, 71.7°), spec. plane (-3-15)

Fig. 5 Experimental EBSD pattern obtained from the lamellar phase indexed as Al_4Sr (tetragonal crystal structure), Mean angle of deviation (MAD) = 0.59°, orientation = (151.4, 93.8, 11.5°), spec. plane (-5-10)

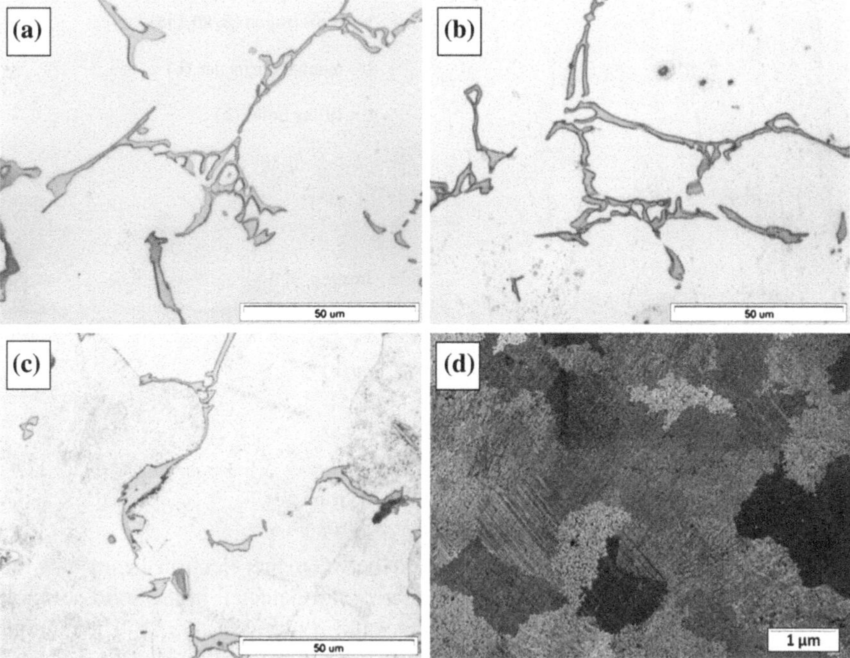

Fig. 6 Examples of revealed microstructures after etching with different reagents: 25 ml H_2O, 75 ml ethylene glycol, 1 ml HNO_3 in bright field, low quality for detection of phases (**a**); 5–20 ml acetic acid, 80–95 ml H_2O in bright field, good quality for detection phases (**b**); 4.2 g picric acid, 10 ml acetic acid, 10 ml H_2O, 70 ml ethanol in bright field, very low quality for detection of phases (**c**); 4.2 g picric acid, 10 ml acetic acid, 10 ml H_2O, 70 ml ethanol in polarized field with quarter-wave plate, good quality for detection of grain boundaries (**d**)

this eutectic phase is composed of mainly aluminium, calcium and magnesium (point 1, Table 2). The second one was a thin lamellar eutectic phase containing mainly magnesium, aluminium and strontium (point 2, Table 2). The third one was globular particles composed mainly of aluminium and manganese (point 3, Table 2). It is worth noting that precise determination of magnesium content in the

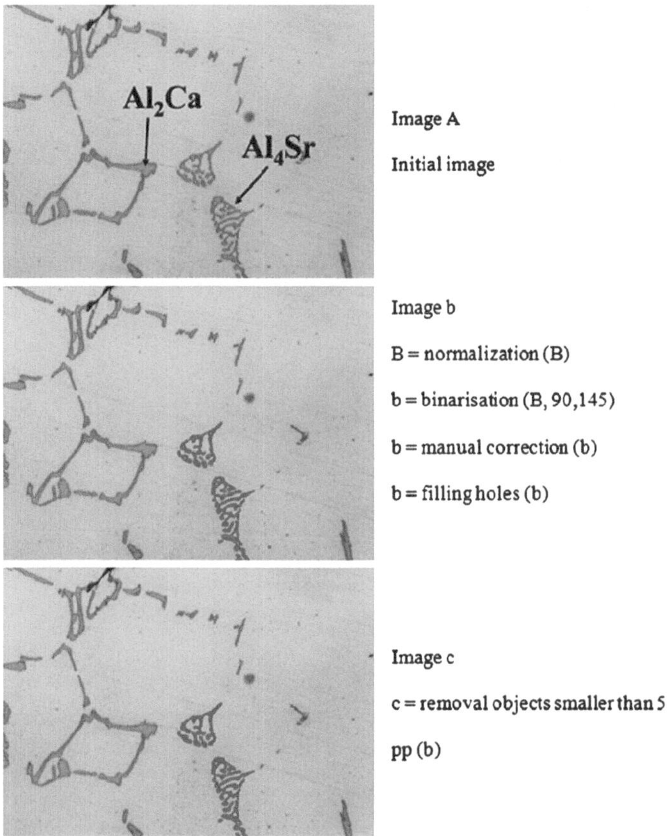

Fig. 7 Semi-automatic procedure used to detect the Al_4Sr phase

second phase is difficult, due to interaction between the electron beam and the magnesium matrix. Calcium and strontium were not detected in the solid solution of α-Mg due to their low solubility in magnesium (point 4, Table 2). The content of aluminum dissolved in α-Mg is higher than its maximal solid solubility at room temperature. It is connected with high cooling rate during the solidification of ingot.

In order to identify the crystalline structure of these intermetallic compounds existing in the as-cast microstructure of the alloy studied, X-ray diffraction analysis was carried out and the results are shown in Fig. 3. It can be seen that the alloy consists of the α-Mg matrix, Al_2Ca and Al_4Sr compounds. Moreover, a few diffraction lines with low intensity were observed which were not positively identified. Presumably, these peaks are produced by particles containing aluminum and manganese (Al_xMn_y phase). In addition, the presence of Al_2Ca and Al_4Sr phases was confirmed by electron back-scattering diffraction (EBSD) analysis as shown in Figs. 4 and 5, respectively.

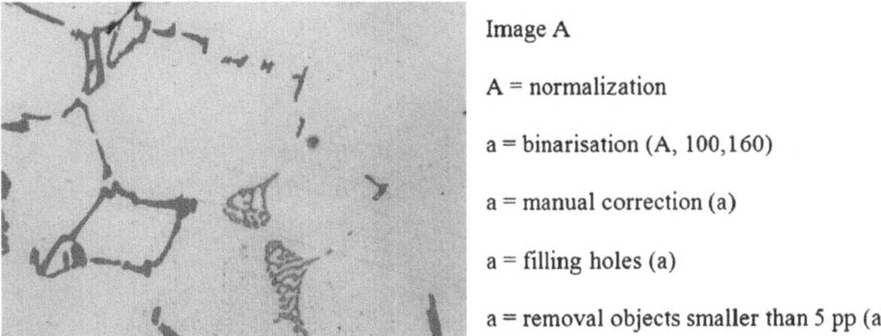

Fig. 8 Semi-automatic procedure used to detect the Al_2Ca phase

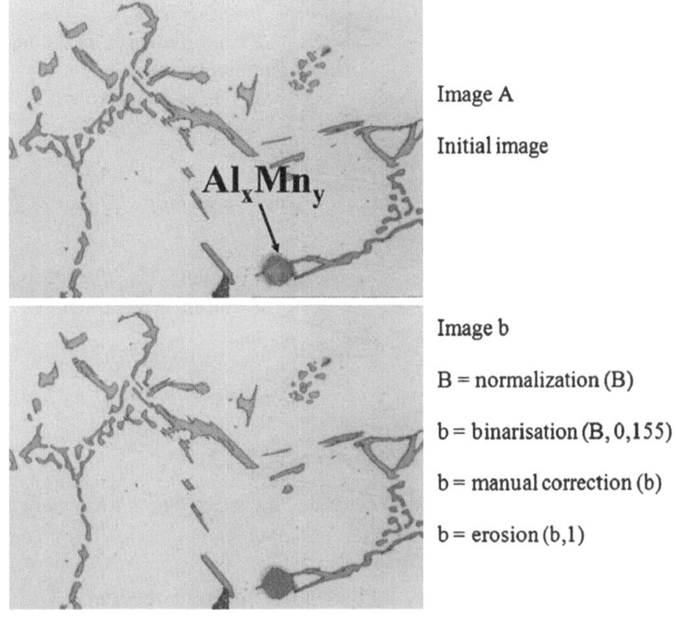

Fig. 9 Semi-automatic procedure used to detect the Al_xMn_y compound

3.2 *Procedures for Detection of Intermetallic Phases and Grain Boundaries*

The selection of the best reagent to reveal the microstructural constituents is an important factor during the development of complex procedures for the quantitative description of the microstructure.

Images of the microstructures after etching in different reagents were registered using the Olympus GX-70 light microscope equipped with a DP70 colour digital camera employing various acquisition techniques. Figure 6 shows examples of

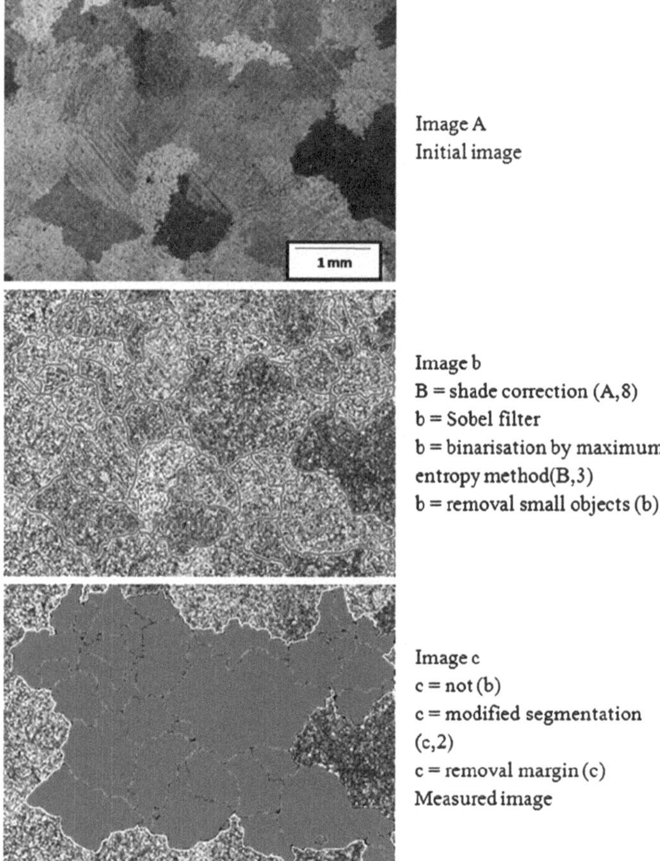

Fig. 10 Semi-automatic procedure used to determine the stereological parameters of grains in MRI-230D alloy

microstructures of MRI230D alloy etched with different reagents. The condition for proper detection is a homogeneous grey level from the objects to be measured, which is different from the background level which is not elevated. An initial description of all images obtained during these investigations suggested that this condition could be fulfilled by structural images obtained for unetched samples (Fig. 1). Usage of reagent containing 5–20 ml acetic acid, 80–95 ml H_2O (Fig. 6b) also allows the correct detection of intermetallic phases, however, the distinction of the compounds is more difficult than in unetched samples.

Since the intermetallic compounds are present in interdendritic regions, distinguishing the grain size from the dendritic cell size in cast magnesium alloys can be challenging. Etching with reagents containing 4.2 g picric acid, 10 ml acetic acid, 10 ml H_2O and 70 ml ethanol and usage of polarized light permits

Table 3 Results of quantitative analysis in MRI-230D magnesium alloy

Parameters	Phase		
	Al_2Ca	Al_4Sr	Al_xMn_y
A_A (%)	6.1	1.5	0.6
$v(A_A)$	35.1	47.3	105
ξ	0.43	0.52	0.86

identification of individual grains and reveals that not all of intermetallic-rich regions correspond to grain boundaries.

Based on the etching results, semi-automatic procedures using the Met-Ilo image analysis program were developed to detect intermetallic phases and grain boundaries in the investigated alloy (Figs. 7, 8, 9, and 10). In these figures, grey images are designated in capital letters and binary images with small letters. Letters in the brackets indicate the analyzed image, the digit means value of step for the image transformation that was used in the Met-Ilo program [6].

When the light microscopy is used to estimate the phase content in the MRI-230D alloy, the most appropriate magnification for correct detection of second phases in MRI-230D alloy is 500x. In order to obtain the accuracy of the measurements of volume fraction of intermetallic phases lower than 3 %, the minimum number of registered images at magnification of 500x is 22. In case of measurements of grain size, repeatable results are obtained if a minimum of 300 grains are analyzed.

The results of the quantitative analysis are shown in Table 3. It can be seen that the dominant intermetallic compound is the Al_2Ca Laves phase. The low values of the variation coefficient of area fraction $v(A_A)$ indicates that Al_2Ca and Al_4Sr phases are a homogenous arrangement in the microstructure. The low values of shape factor ξ of these compounds correspond to their irregular shape. In the case of the Al_xMn_y phase, the high value of shape factor indicates on near-globular morphology. The grain size of the ingot analyzed is 424 μm, while the dendritic cell size is only 114 μm.

4 Conclusion

The as-cast microstructure of MRI-230D alloy consists of dendritic supersaturated α-Mg matrix, coarse irregular-shaped eutectic α-Mg+Al_2Ca, thin lamellar eutectic α-Mg+Al_4Sr distributed in the interdendritic regions and globular particles of Al_xMn_y compound inside the grains. The dominant intermetallic compound is Al_2Ca and its area fraction is 6.1 %. The performed investigations have shown that the detection of the particles occurring in the MRI-230D magnesium alloy can be performed on unetched samples. Difference in grey level between precipitates is sufficient for detection of Al_2Ca, Al_4Sr and Al_xMn_y phases. Therefore, a simple semi-automatic procedure using the Met-Ilo image analysis program can be applied to estimate the phase content and grain size in the MRI-230D alloy. These procedures can be used in the control of quality of ingots, sand and squeeze casts.

Acknowledgments The present work was supported by the Polish Ministry of Science and Higher Education under the strategical project no. POIG.01.01.02-00-015/09 (FSB-71/RM3/2010).

References

1. Aghion, E., Lulu, N., Moscovitch, N., Bronfin, B.: The environmental behavior of magnesium alloy MRI-230D. In: Proceedings Magnesium of the 8th International Conference on Magnesium Alloys and their Applications, Weimar, pp. 887–892 (2009)
2. Aljarrah, M., Medraj, M., Jian, Li., Essadiqi, E.: Phase equilibria of the constituent ternaries of the Mg–Al–Ca–Sr system. JOM 68–74 (2009)
3. Buehler Ltd.: Buehler's Guide to Materials Preparation. Illinois, USA (2001)
4. Janz, A., Gröbner, J., Schmid-Fetzer, R.: Thermodynamics and constitution of Mg–Al–Ca–Sr–Mn alloys: Part II J. Phase Equilib. Diffus. **30**, 157–175 (2009)
5. Luo, A.A., Balogh, M.P., Powell, B.R.: Creep and microstructure of magnesium–aluminum–calcium based alloys. Met. Mat. Trans. A **33**, 567–574 (2002)
6. Szala, J.: Application of computer-aided image analysis methods for a quantitative evaluation of material structure (in Polish). Silesian University of Technology, Gliwice (2001)
7. Terada, Y., Itoh, D., Sato, T.: Creep rupture properties of die-cast Mg–Al–Ca alloys. Mat. Chem. Phys. **113**, 503–506 (2009)
8. Wang, Y., Guan, S., Zeng, X., Ding, W.: Effects of RE on the microstructure and mechanical properties of Mg8Zn4Al magnesium alloy. Mat. Sci. Eng. A **416**, 109–118 (2006)
9. Xu, S.W., Matsumoto, N., Yamamoto, K., et al.: High temperature tensile properties of—as-cast Mg–Al–Ca alloys. Mat. Sci. Eng. A **509**, 105–110 (2009)
10. Zhu, S.M., Mordike, B.L., Nie, J.F.: Creep properties of a Mg–Al–Ca alloy produced by different casting technologies. Mat. Sci. Eng. A **483–484**, 583–586 (2008)

Dynamic Analysis of Pre-Cast RC Telecommunication Towers Using a Simplified Model

Marcelo A. Silva, Jasbir S. Arora and Reyolando M. L. R. F. Brasil

Abstract The main goal of this chapter is to propose a simplified procedure to accomplish the dynamic analysis of pre-cast RC (reinforced concrete) telecommunication towers subjected to wind loads. With the methodology proposed here, only the static results and the first natural frequency of vibration are needed to accomplish the dynamic analysis of a given structure. The method is easier and faster than the traditional dynamic analysis approach. In this work, results of the dynamic and static analysis of 90 real structures are used in the optimization process. The difference between the results given by the simplified method proposed here and the complete dynamic analysis are less than 2 %.

1 Introduction

The theme of the present work is related to the recent public auctions for operation of cellular telephony in Brazil, in line with the natural growth of the sector of telecommunication in the world. With this, new systems of transmission and

M. A. Silva (✉)
RM Soluções Engenharia Ltda, Av. General MacArthur, 330,
2° Andar, São Paulo-SP CEP 05.338-000, Brazil
e-mail: marcelo@rmsol.com.br

J. S. Arora
The University of Iowa, Virtual Soldier Research (VSR) Program/CCAD,
4110 SC/Eng/UIowa, Iowa City, IA 52242, USA
e-mail: jasbir-arora@uiowa.edu

R. M. L. R. F. Brasil
Department of Structural and Geotechnical Engineering-Polytechnic School, The University of Sao Paulo, Cx. Postal 61548São Paulo-SP CEP 05.424-970, Brazil
e-mail: rmlrdfbr@usp.br

A. Öchsner et al. (eds.), *Design and Analysis of Materials and Engineering Structures*, Advanced Structured Materials 32, DOI: 10.1007/978-3-642-32295-2_8,

Fig. 1 Typical pre-cast RC telecommunication tower

reception of electromagnetic waves are being installed. Due to the restrictions of the local legislations, the installation of new towers has been an extremely difficult work. One possible solution is the sharing of existing structures, where several telecommunication companies use a same tower already in place. It is necessary to verify if the existing structure, when subjected to the loads of a given configuration of antennas and supports, presents appropriate safety. The installation of the systems must be accomplished in very few months. Such fact requires the analysis, with reliability and speed, of thousands of existing telecommunications towers in a short period of time. A typical structure analyzed here is shown in Fig. 1.

Unfortunately, some accidents occurred with some of these structures. In Fig. 2, we show one structure collapsed during a wind storm. Among several hypothesis, based on investigations, studies and tests, the main causes of the collapses are:

- design error in the junction of the sections;
- over load; the number of antennas, platforms and supports are larger than the prescribed one; the place of installation of the tower is different from the one considered in the design;
- structures designed without considering the dynamic effects of the wind loads.

Fig. 2 Pre-cast RC structure collapsed during a wind storm

The present work deals with the dynamic effects of the wind loads. We show here that the dynamic effects of wind are very significant and can drastically contribute to structural collapse. The main feature researched is the dynamic magnification factor, defined here as the ratio between the bending moment given by the dynamic and static models [1]. Surfaces are created to give this factor as a function of the structure height and the first natural frequency of vibration. To create these surfaces, optimization problems were formulated where the objective function is the error between the dynamic magnification factor, computed according to NBR-6123 [1], and other given by equations, defined as a function of the structure height and first natural frequency of vibration. The design variables are the coefficients of these equations, and constraints are imposed to avoid a negative magnification factor. The static analysis is quite straightforward and easy to implement. However, the same thing is not true for the dynamic analysis because it requires computation of natural frequencies and mode shapes and coefficient of amplification, beside other variables. The main goal of this chapter is to define a procedure to simplify the dynamic analysis of Reinforced Concrete (RC) telecommunication towers. Using graphs created by the authors, engineers can easily obtain the dynamic response of a tower only by multiplying the results of the static analysis by the coefficients given in graphs. In this kind of structures, the main internal loads are the bending moment, the shear load and the axial load. Only the flexure moment and the shear load are multiplied by the magnification factor to obtain the dynamic results. In cases analyzed here, the axial load is not influenced by the dynamic response.

To solve the problems discussed here, use of advanced tools and models is needed, such as the reinforced concrete modeling, dynamic analysis of structures subjected to wind loading and Lagrangian optimization. First the linear dynamic model used to accomplish the structural analysis is presented. This model is based on the discrete dynamic model of the NBR-6123 code [1]. In this model, the structure's effective stiffness is considered constant computed as the gross moment of inertia. Next, the optimization problem is described, containing a description of the structures, the formulation of the problem and the results obtained. Finally, conclusions based on the present study and suggestions for further works are presented.

2 Review of Literature

The RC analysis is done based on the NBR-6118 code [2]. A review of the literature indicates that the effective stiffness of RC structures depends on the bending moment as well as the distribution and the amount of reinforcement. An equation proposed by Branson [5] for the calculation of the effective stiffness was incorporated in ACI-318 [3] and, more recently, in NBR-6118 [2]. Several researchers have used Branson's equation to compute the displacement of RC beams and slabs. Inspired by Branson's equation, (Brasil and Silva 2006) present a methodology for calculation of effective stiffness of RC beams subjected to bending moment and shear load. The methodology consists of using optimization techniques to minimize the error between the displacements measured in tests with those given by the integration of the elastic line, thus representing the effective stiffness of the member. In the present work, according to the dynamic linear model, we consider constant stiffness as the gross section moment of inertia.

In the work of Brasil and Silva [6], a non-linear dynamic model is presented for analysis of slender RC structures under dynamic wind loading. The model is based on ABNT [1] and on the effective stiffness equations given by Silva and Brasil [6]. Once equations for the calculation of the effective stiffness are adopted, a non-linear static analysis of the structure under the mean wind speed loading is accomplished first [1] where, in each iteration of the P-delta method, the effective stiffness of RC elements is computed as a function of the bending moment. Considering the effective stiffness obtained in the final iteration of the P-delta method, the authors calculate the natural frequencies and modes of vibration of the structure. In this kind of problem, once the moment given by the P-delta method is greater than the linear static one, the effective stiffness may be lower than estimated by the linear static analysis, and lower frequencies will be generated. These modes and frequencies are then used to perform the floating wind analysis of the structure [1]; that is the analysis of the vibration of the structure due to the variation of the wind velocity. These authors consider the structure to vibrate under the wind loading around the equilibrium configuration given by the P-delta method, and that the amplitude of displacement is given by the dynamic wind analysis. The sum of the non-linear static analysis and the floating wind analysis results constitutes the non-linear dynamic analysis of the structure. This process is depicted in Figs. 3 and 4. Silva and Brasil [6] conclude that the dynamic internal loads of the non-linear dynamic model are 15 % larger than those with the linear dynamic model [1]. When the values obtained with the non-linear dynamic model are compared with those given by the linear static model of ABNT [1], it is concluded that they are 50 % larger than the linear

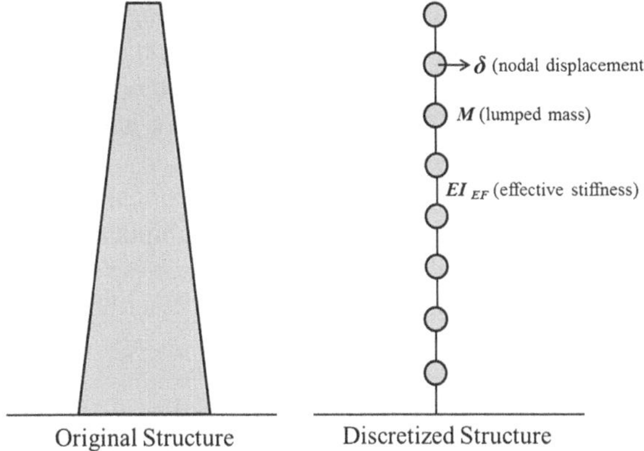

Fig. 3 Original and discretized structure

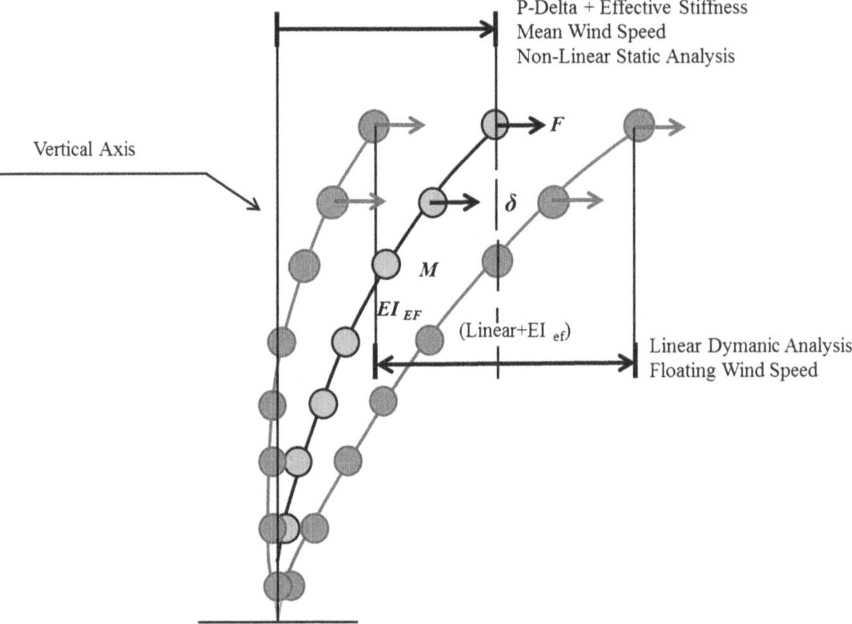

Fig. 4 Graphical representation—non-linear dynamic model [8]

static analysis results. In the present work, the authors propose a code based method, therefore, only the linear dynamic analysis is considered. It is done by taking the effective stiffness constant as the gross section moment of inertia and neglecting second order effects (P-delta effects).

For the optimization process, the augmented Lagrangian method for dynamic response optimization problems as described by Chahande and Arora [7] and Arora [4] is used. This method transforms a constrained optimization problem into an unconstrained optimization problem. The objective and constraints functions are combined using the Lagrange multipliers and penalty parameters to create an augmented Lagrangian functional. A sequence of functionals is created by properly altering the penalty parameters and Lagrange multipliers. The unconstrained minimum value of the functional in this sequence converges to the minimum of the constrained problem. Considering all the methods described in this section, we developed the work here presented.

3 Linear Dynamic Analysis

3.1 Linear Static Analysis

According to ABNT [1], V_0 (m/s) is the mean wind speed computed based on a 3 s interval, at 10 m above ground, for a plain terrain with no roughness, and a return period of 50 years. The topographic factor is S_1, while the terrain roughness factor is S_2, given as

$$S_2 = bF_r(z/10)^p \tag{1}$$

where b, p and F_r are factors that depend on the terrain characteristics and z is the height above ground in meters. Factors S_1, S_2 and S_3 are given in tables in ABNT [1]. The characteristic wind speed (m/s) and the wind pressure (Pa), respectively, are

$$V_k = V_0 S_1 S_2 S_3 \quad \text{and} \quad q = 0.613 V_k^2 \tag{2}$$

The wind load (N) on an area A (projected area on a vertical plane of the given object in m^2) is computed as

$$F = C_a A q \tag{3}$$

where C_a is the drag coefficient. ABNT [1] presents tables for C_a values.

3.2 Linear Dynamic Analysis

If the first natural frequency of vibration of a given structure is smaller than 1 Hz, it is necessary to perform a dynamic analysis of the structure [1]. According to ABNT [1], the dynamic analysis is performed as follows. For the jth degree of freedom, the total load X_j due to direct wind along the tower is the sum of the mean and the floating loads given as:

$$X_j = \bar{X}_j + \hat{X}_j \tag{4}$$

The mean load $\bar{X}_j$ is given as

$$\bar{X}_j = \bar{q}_o b^2 C_j A_j \left(\frac{z_j}{z_r}\right)^{2p} \tag{5}$$

where

$$\bar{q}_o = 0.613\bar{V}_p^2 \quad \text{and} \quad \bar{V}_p = 0.69V_0S_1S_3 \quad (\bar{q}_o \text{in N/m}^2 \text{ and } \bar{V}_p \text{in m/s}) \tag{6}$$

and b and p are given in Table 20 in ABNT [1]; z_r is the reference height, taken as 10 m in this work; $\bar{V}_p$ is the design wind speed corresponding to the mean speed during 10 min at 10 m above ground for a terrain roughness (S_2) for category II.

The floating component $\hat{X}_j$ in Eq. (4) is given as

$$\hat{X}_j = F_H \psi_j \varphi_j \tag{7}$$

where

$$\psi_j = \frac{m_j}{m_o}, \quad F_H = \bar{q}_o b^2 A_o \frac{\sum_{i=1}^{n} \beta_i \varphi_i}{\sum_{i=1}^{n} \psi_i \varphi_i^2} \xi, \quad \beta_i = C_{ai} \frac{A_i}{A_o} \left(\frac{z_i}{z_r}\right)^p \tag{8}$$

and m_i, m_0, A_i, A_0, ξ and C_{ai} are the lumped mass at the ith degree of freedom, a reference mass, the equivalent area at the ith degree of freedom, a reference area, the dynamic coefficient of amplification in Figs. 14 to 18 of ABNT [1], and the drag coefficient for area A_i, respectively. Note that $\boldsymbol{\varphi} = [\varphi_i]$ is a given mode of vibration. To compute φ_i and ξ, it is necessary to consider the structural mass and stiffness. The damping factor ζ is given by ABNT [1]. When the section varies as a function of the diameter $\zeta = 1.5$ % and for cylindrical structures $\zeta = 1.0$ %. The lumped mass can be easily calculated by summing the mass around the influence region of a node. The total homogenized moment of inertia of the cross-section is given as

$$I_{\text{total}} = I_G + I_{\text{s hom}}, \quad I_{\text{s hom}} = I_{\text{s}} \left(\frac{E_{\text{s}}}{E_{\text{csec}}} - 1\right), \quad E_{\text{csec}} = 0.85 \times 5600\sqrt{f_{\text{ck}}} \text{ (MPa)}, \tag{9}$$

where E_{s}, $E_{\text{c sec}}$, I_{s}, $I_{\text{s hom}}$, I_G and f_{ck} are the elasticity modulus of steel, the secant elasticity modulus of concrete [2], the moment of inertia of the total longitudinal steel area, the homogenized moment of inertia of the longitudinal steel area, the gross moment of inertia of the cross-section, and the characteristic compressive resistance in MPa for 28 days old concrete, respectively. In the linear dynamic analysis here accomplished, we consider the cross-sectional moment of inertia as,

$$I_{EF} = I_G, \tag{10}$$

of each section to compute the stiffness matrix of the structure. We consider that the structure under linear static behavior does not suffer any plastification or irreversible cracking.

Consider a given vector $\hat{Q}_i$ which represents a quantity such as internal loads, stress, or strain, due to the ith natural mode of vibration. The contribution $\hat{Q}$ of r modes in the dynamic analysis is computed as

$$\hat{Q} = \left[\sum_{k=1}^{r} \hat{Q}_i^2\right]^{1/2}, \quad \text{while} \quad Y_i = \frac{1}{3}X_i \tag{11}$$

is the transverse load, due to the variation of the wind direction. The first expression of Eq. (11) is called simply the *RSS method, which means square root of the sum* of the *squares method.*

4 The Augmented Lagrangian Method

To solve the structural optimization problem formulated in this chapter, it is necessary to adopt optimization algorithms that deal with static and dynamic constraints, as well as non-linear functions. A non-linear optimization problem with static and dynamic constraints is presented as follows: find the design variables $\boldsymbol{b} \in \boldsymbol{R}^n$ that minimize the non-linear objective function $f(\boldsymbol{b})$ subjected to the non-linear constraint functions:

Static

$$g_i(\boldsymbol{b}) = 0; \quad i = 1, l \tag{12}$$

$$g_i(\boldsymbol{b}) \leq 0; \quad i = l + 1, m \tag{13}$$

and dynamic, $\forall\, t \in [t_0, t_f]$,

$$g_i(\boldsymbol{b}, t) = g_i(\boldsymbol{b}, z(t), \dot{\mathbf{z}}(t), \ddot{\mathbf{z}}(t), t) = 0; \quad i = m + 1, l' \tag{14}$$

$$g_i(\boldsymbol{b}, t) = g_i(\boldsymbol{b}, z(t), \dot{\mathbf{z}}(t), \ddot{\mathbf{z}}(t), t) \leq 0; \quad i = l' + 1, m' \tag{15}$$

In dynamic of structures, the displacement vector $z(t)$ must satisfy the equations of motion, a system of second-order ordinary differential equations:

$$\boldsymbol{M}(\boldsymbol{b}, \ddot{\mathbf{z}}, \dot{\mathbf{z}}, \mathbf{z}, (t)\ddot{\mathbf{z}}(t) + \boldsymbol{C}(\boldsymbol{b}, \ddot{\mathbf{z}}, \dot{\mathbf{z}}, \mathbf{z}, t)\dot{\mathbf{z}}(t) + R(\boldsymbol{b}, \ddot{\mathbf{z}}, \dot{\mathbf{z}}, \mathbf{z}, t) = p(\boldsymbol{b}, \ddot{\mathbf{z}}, \dot{\mathbf{z}}, \mathbf{z}, t) \tag{16}$$

with the initial conditions as $\dot{\mathbf{z}}(t_0) = \dot{\mathbf{z}}_0$ and $\mathbf{z}(\mathbf{t_0}) = \mathbf{z_0}$. Here $\boldsymbol{M}$ $(\boldsymbol{b}, \ddot{\mathbf{z}}, \dot{\mathbf{z}}, \mathbf{z}, t)$ and $\boldsymbol{C}$ $(\boldsymbol{b}, \ddot{\mathbf{z}}, \dot{\mathbf{z}}, \mathbf{z}, t)$ are the mass and damping matrices, respectively; the vector $\boldsymbol{R}$ $(\boldsymbol{b}, \ddot{\mathbf{z}}, \dot{\mathbf{z}}, \mathbf{z}, t)$ is the generalized elastic force; and $\boldsymbol{p}$ $(\boldsymbol{b}, \ddot{\mathbf{z}}, \dot{\mathbf{z}}, \mathbf{z}, t)$ is the generalized force vector. Equations (12) and (13) include, for example, limits of the design variables. Equations (14) and (15) represent constraints on dynamic response such as the maximum and minimum values of the displacements $z(t)$, dynamic strain and stress. The initial and final times are t_0 and t_f, respectively.

During the design process, a set of parameters, denoted as design variables $\boldsymbol{b}$, are selected to define the system. Once the design variables are specified, the vector $z(t)$ is determined by Eq. (16). The displacements are called state variables, as they determine the structure configuration for every $t \in [t_0, t_f]$, and Eq. (16) is the state equation. The constraints on the dynamic response are explicit functions of state variables and implicit functions of the design variables. Usually, in structural optimization problems with dynamic constraints, the state variables vector is denoted as $z(t)$. In this work, we use an equivalent static method where the analysis is done by computing the maximum amplitude of each mode of vibration and then using the *RSS* method, as shown previously in Eq. (11).

There are several optimization methods to solve the problem defined by Eqs. (12)–(16). We treat this problem using the augmented Lagrangian method where an augmented functional is created using the objective and the constraint functions:

$$\Phi(\boldsymbol{b},\ \boldsymbol{u},\ \boldsymbol{r}) = \boldsymbol{f}(\boldsymbol{b}) + \boldsymbol{P}(\boldsymbol{g}(\boldsymbol{b}),\ \boldsymbol{u},\ \boldsymbol{r}) \tag{17}$$

$P(\boldsymbol{g}(\boldsymbol{b}),\ \boldsymbol{u},\ \boldsymbol{r})$ is a penalty functional, and $\boldsymbol{u} \in \boldsymbol{R}^{m'}$ and $\boldsymbol{r} \in \boldsymbol{R}^{m'}$ are the Lagrange multipliers and the penalty parameters, respectively. It is possible to determine $\boldsymbol{u}^*$ and $\boldsymbol{r}^*$ so that the minimum point $\boldsymbol{b}^*$ of the functional defined in Eq. (17) is the minimum point of the problem defined by Eqs. (12)–(16). As we use iterative methods to find $\boldsymbol{u}^*$ and $\boldsymbol{r}^*$, it is necessary to use a stopping criterion.

The augmented Lagrangian method defines procedures for updating penalty parameters and Lagrange multipliers. This method can be simply described by the algorithm:

Step1: Set $k = 0$, estimate vector u and r.
Step2: Minimize $\Phi(b,u^k,r^k)$ with respect to b. Let b^k be the best point obtained in this step.
Step3: If the stopping criterion is satisfied, stop the iterative process.
Step4: Update u^k and r^k if necessary.
Step5: Set $k = k + 1$ and go to Step 2.

The multipliers method is quite simple, and its essence is contained in Steps 2 and 4. The functional in Eq. (17) can be defined in several ways. In the present work, the Lagrangian functional adopted to solve the problem with dynamic response is defined as:

$$\begin{aligned}\Phi(\boldsymbol{b},\ \boldsymbol{\theta},\ \boldsymbol{r}) = \boldsymbol{f}(\boldsymbol{b}) &+ \frac{1}{2}\left\{\sum_{i=1}^{l} r_i(g_i(\boldsymbol{b}) + \boldsymbol{\theta}_i)^2 + \sum_{i=l+1}^{m} r_i(g_i(\boldsymbol{b}) + \boldsymbol{\theta}_i)_+^2\right\}\\ &+ \frac{1}{2}\int_{t_0}^{t_f}\left\{\sum_{i=m+1}^{l'} r_i(g_i(\boldsymbol{b},\boldsymbol{t}) + \boldsymbol{\theta}_i)^2 + \sum_{i=l'+1}^{m'} r_i(g_i(\boldsymbol{b},\ \boldsymbol{t}) + \boldsymbol{\theta}_i)_+^2\right\}dt\end{aligned} \tag{18}$$

In Eq. (18) r_i represents the penalty parameters and θ_i defines the Lagrangian multipliers as $u_i = r_i\theta_i$, for $i = 1,\ldots,m'$. Note that, in Eq. (18), the dynamic constraint functions are integrated over the time interval and combined with the objective function to obtain the Lagrangian functional. At Steps 2 and 3, it is necessary to adopt a stopping criterion. Here, we consider the following stopping criterion:

$$K < p, \tag{19}$$

where p is the maximum number of iterations, and

$$\left\|\nabla\Phi(\boldsymbol{b}^k, \boldsymbol{u}^k, \boldsymbol{r}^k)\right\| \leq \varepsilon \quad \text{and} \tag{20}$$

$$\begin{gathered} K_b = \max\{\max_{1\leq i\leq l}|g_i(\boldsymbol{b}^{(k)})|;\ \max_{l+1\leq i\leq m}|\max(g_i(\boldsymbol{b}^{(k)}), -\theta_i)|; \\ \max_{1\leq i\leq l'}(\max_{t_0\leq t\leq t_f}|g_i(\boldsymbol{b}^{(k)}, t)|);\ \max_{l'+1\leq i\leq m'}(\max_{t_0\leq t\leq t_f}|\max(g_i(\boldsymbol{b}^{(k)}, t), -\theta_i(t))|)\} \leq \varepsilon \end{gathered} \tag{21}$$

$$\left\|\left[\Phi(\boldsymbol{b}^{k+1}, \boldsymbol{u}^{k+1}, \boldsymbol{r}^{k+1}) - \Phi(\boldsymbol{b}^k, \boldsymbol{u}^k, \boldsymbol{r}^k)\right]/\Phi(\boldsymbol{b}^k, \boldsymbol{u}^k, \boldsymbol{r}^k)\right\| \leq \varepsilon \tag{22}$$

In Eq. (21), K_b is the maximum constraint violation, and in Eqs. (20)–(22) ε is the tolerance. If the algorithm is not converging, the condition in Eq. (19) states a finite number of iterations. Chahande and Arora [7] noted in several examples analyzed that the best value for p is $2n$. The process of updating the Lagrange multipliers and penalty parameters and more details on the adopted algorithm are presented by Arora [4] and Chahande and Arora [7]. In the work of Chahande and Arora [7], only Eq. (21) is suggested as the stopping criterion, but in this work, based on tests accomplished by the authors, we adopt other conditions as well.

A computer program was developed, based on the augmented Lagrangian method. The following numerical methods were utilized to implement this method:

- to solve the equations of motion (16), we use the procedure given in Sect. 3.2 for dynamic linear analysis;.
- regarding the unconstrained minimization (Step 2), we use the conjugate gradient method with Armijo line search;
- to calculate the gradient vector, we use the finite difference method;
- to solve linear systems, we use Cholesky decomposition.

5 Description of the Structures

The RC towers analyzed here present a height varying from 20 to 60 m with a circular cross-section. Based on that, we suggest that the results given here be used only for structures with height no larger than 60 m. The diameter, thickness and steel areas may change along the height of the tower and are different from one

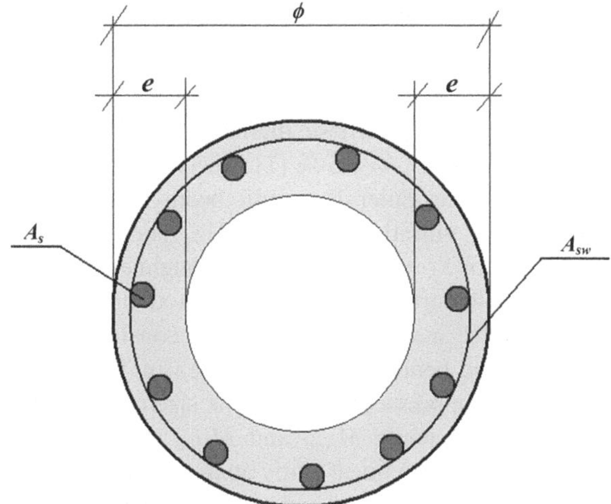

Fig. 5 RC telecommunication tower cross-section

tower to the other. The concrete used in the fabrication of the tower has a characteristic resistance at 28 days (f_{ck}) as 45 MPa, which gives a secant elasticity modulus (E_{csec}) of 31.9 GPa from Eq. (9). Applying the safety factor, the concrete design resistance is then $f_{cd} = 45/1.3$ MPa. The steel has a covering of 25 mm, design resistance of $f_{yd} = 500/1.15$ MPa and elasticity modulus $E_s = 210$ GPa.

The structures were discretized with 41 nodes and 40 elements. The first element starts at the first node and ends at the second, the second element starts at the second node and ends at the third, and so on. With this discretization, the structure has 240 degrees of freedom. The displacement vector corresponding to the structural degrees of freedom is also denoted as the state variable vector. Figure 5 shows a typical structure cross-section, where ϕ, e, A_s and A_{sw} are respectively the external diameter, thickness, the longitudinal steel area and the traverse reinforcement.

As the ninety structures analyzed here are installed in several places in Brazil, the basic wind speed V_0, topographic factor S_1 and terrain roughness $S_2 \equiv (b; p; F_r)$, are adopted according to the local of installation. The statistical factor is $S_3 = 1.1$. As stated before, the wind load on an area A is $F = C_a A q$, where C_a represents the drag coefficients. Several pieces of equipment are installed on the structure, such as a ladder with anti-fall cable, a platform with antenna supports, night signer lights, a system of protection against lightning, and antennas. The values of A and C_a change from tower to tower.

Besides these areas and drag coefficients, the tower mass is considered distributed along the structure proportionally to the volume and can be computed using a density of 2,500 kg/m^3. The masses of the other components are also considered. As the goal of the chapter is to compare the static with the dynamic internal loads, specially the bending moment, these data are not important to the final result of the work.

6 Computation of the Dynamic Magnification Factor

To explain the methodology and the results obtained here, let us take as example a 40 m tall structure, whose diameter varies along the height. In this case, the damping factor is $\zeta = 1.5\ \%$ [1]. Accomplishing a dynamic finite element analysis and considering Euler–Bernoulli beam elements, we compute the first natural frequency of vibration $f_1 = 0.42$ Hz. The wind factors are $V_0 = 40$ m/s, topographic factor $S_1 = 1.0$, and terrain roughness $S_2 \equiv$ III and the statistical factor is $S_3 = 1.1$. The structure is loaded with antennas, supports, platforms, cables, ladder, etc. After analyzing this structure considering the static and dynamic models, we obtain the bending moments given in Table 1. Note that the results of some nodes were suppressed to reduce the size of the table. In this Table, z is the height above ground, while M_{lsad} and M_{ldad} are the design bending moments given, respectively, by the linear static analysis and linear dynamic analysis. We consider the bending moment to develop the methodology proposed here because it is the most important internal load in this kind of structure. The magnification factor $\gamma_d(i)$ of a given section (i) is computed as

$$\gamma_d(i) = \frac{M_{ldad}(i)}{M_{lsad}(i)} \tag{23}$$

The magnification factor of the structure γ_d is the average of $\gamma_d(i)$ computed in the results of nodes 27–41 (approximately the one-third of the structure near to the ground level):

$$\gamma_d = \frac{1}{15}\sum_{i=27}^{41} \gamma_d(i) \tag{24}$$

In Table 1, in the resume line, 100 % means the result of the static analysis while 138 % is the ratio between the dynamic and static results. It means that the dynamic flexure moment is 38 % larger than the static flexure moment (average in the third part of the structure near to the ground level). Therefore, γ_d, defined here as the dynamic magnification factor, is the factor that can be used to find directly the dynamic results of a given tower. Figure 6 shows these flexure moments along the structure axis. Adopting the same procedure for several structures and grouping those that present the same value of ζ and S_2, we can build other tables, similar to Table 2. In this table H (m) is the structure height, f_1 (Hz) is the first natural frequency of vibration and γ_d the dynamic magnification factor.

Consider now that γ_d is written approximately as:

$$\gamma_d \cong \gamma_{dk}, \text{ where } \begin{cases} k = 0 \rightarrow \gamma_{d0} = a \\ k = 1 \rightarrow \gamma_{d1} = a + bH + cf_1 \\ k = 2 \rightarrow \gamma_{d2} = a + bH + cf_1 + dHf_1 + eH^2 + ff_1^2 \end{cases} \tag{25}$$

Table 1 Internal loads of a given 40 m tall structure

Node (i)	Z (m)	Mlsad (kN.m)	Mldad (kN.m)	γ_d
1	40	0.00	0.00	–
2	39	19.40	16.20	0.83
3	38	39.82	35.18	0.88
4	37	61.27	56.84	0.93
10	31	212.00	241.07	1.14
20	21	550.32	730.47	1.33
26	15	807.20	1109.14	1.37
27	14	853.96	1176.92	1.38
28	13	901.85	1245.82	1.38
29	12	950.87	1315.77	1.38
30	11	1000.99	1386.70	1.39
31	10	1052.22	1458.51	1.39
32	9	1104.54	1531.13	1.39
33	8	1157.91	1604.48	1.39
34	7	1212.33	1678.49	1.38
35	6	1267.76	1753.09	1.38
36	5	1324.18	1828.20	1.38
37	4	1381.56	1903.78	1.38
38	3	1439.85	1979.75	1.37
39	2	1499.02	2056.01	1.37
40	1	1558.76	2132.47	1.37
41	0	1619.01	2209.14	1.36
Resume		100 %	138 %	1.38

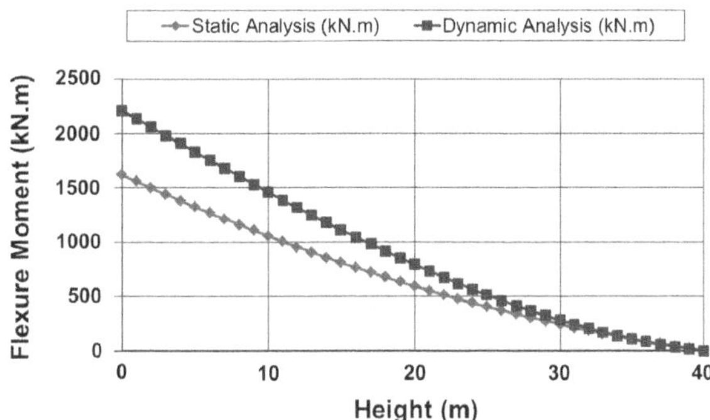

Fig. 6 Dynamic and static flexure moment along the structure axis

In Eq. (25) we note that for $k = 0$ the approximation of γ_d is given by a constant function, when $k = 1$ by a linear function and for $k = 2$ by a quadratic equation.

Considering the approximations given by Eq. (25), we define the following design variables of the optimization problem to determine the dynamic magnification factor:

Table 2 Values of γ_d for $\zeta = 1.5$ % and $S_2 = \text{III}$

H (m)	f_1 (Hz)	γ_d	γ_{d0}	γ_{d1}	γ_{d2}	E_0	E_1	E_2
20	0.97	1.28	1.44	1.28	1.3	0.012	0.000	0.000
30	0.45	1.38	1.44	1.4	1.38	0.002	0.001	0.000
40	0.31	1.42	1.44	1.46	1.45	0.000	0.001	0.001
50	0.2	1.56	1.44	1.51	1.54	0.007	0.000	0.000
60	0.19	1.53	1.44	1.55	1.53	0.004	0.000	0.000
20	1.15	1.27	1.44	1.25	1.27	0.014	0.000	0.000
30	0.54	1.36	1.44	1.38	1.35	0.003	0.000	0.000
40	0.42	1.38	1.44	1.44	1.38	0.002	0.002	0.000
50	0.28	1.5	1.44	1.5	1.47	0.002	0.000	0.000
60	0.23	1.49	1.44	1.54	1.5	0.002	0.001	0.000
20	0.65	1.34	1.44	1.33	1.33	0.004	0.000	0.000
30	0.39	1.42	1.44	1.41	1.4	0.000	0.000	0.000
40	0.26	1.45	1.44	1.46	1.48	0.000	0.000	0.001
50	0.17	1.58	1.44	1.51	1.57	0.012	0.003	0.000
60	0.14	1.59	1.44	1.55	1.59	0.011	0.000	0.000
Total error						0.075	0.009	0.002

$$\boldsymbol{b} = \begin{cases} [a] \\ [a \quad b \quad c] \\ [a \quad b \quad c \quad d \quad e \quad f] \end{cases} \tag{26}$$

Consider respectively the real and approximated dynamic magnification factor vectors $\boldsymbol{\gamma_d}$ and $\boldsymbol{\gamma_{dk}}$ of a given group of structures which present the same ζ and S_2. The quadratic error between the approximation $\boldsymbol{\gamma_{dk}}$ and the real factor $\boldsymbol{\gamma_d}$ is:

$$E_k(\boldsymbol{b}) = \frac{1}{2}\sum_{i=1}^{j} \left(\gamma_{dki}(\boldsymbol{b}) - \gamma_{di}\right)^2 \tag{27}$$

Once γ_{dki} is computed according Eq. (25), we always have an approximation error in (27), the problem of minimizing this error is an optimization problem. We formulate the following optimization problems: determine $\boldsymbol{b}$ that

Minimize $E_k(\boldsymbol{b})$

Subjected to

$$\gamma_{dki}(\boldsymbol{b}) \geq 0 \; i = 1, \ldots, j \tag{28}$$

Note that for $k = 0$ we have one optimization problem, $k = 1$ other, and so on, and that the value of j is 15. The optimization problems presented here are of inverse analysis. As we mentioned in the present work, we analyzed a total of ninety towers and computed the real dynamic magnification factor vectors of groups of these towers with the same ζ and S_2. These vectors were used to compute the approximation dynamic magnification factor (Figs. 7, 8, 9, 10, 11, 12).

Coming back to Table 2, γ_{d0} is the approximation of the constant function, while γ_{d1} and γ_{d2} are the approximations given by the linear and quadratic

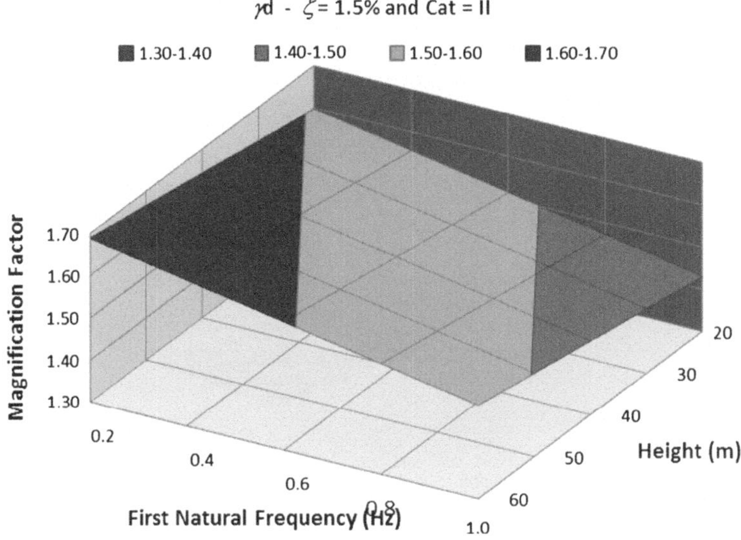

Fig. 7 Dynamic magnification factor γ_{d1} for $\zeta = 1.5\ \%$ and $S_2 = \text{II}$

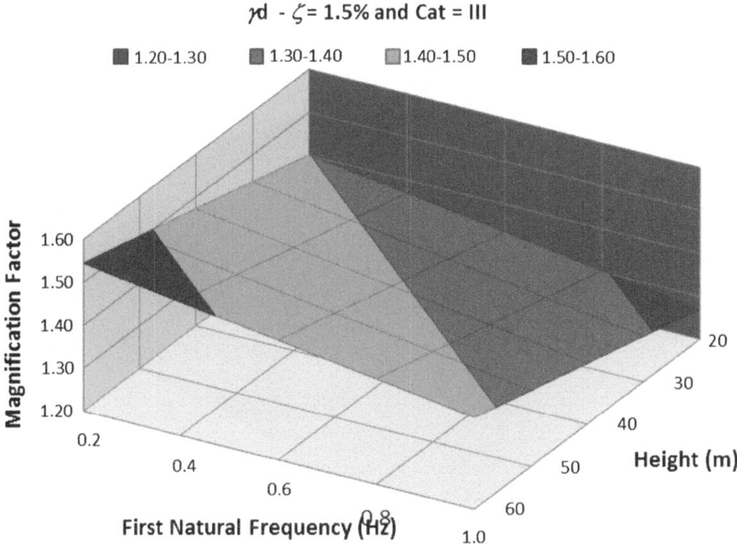

Fig. 8 Dynamic magnification factor γ_{d1} for $\zeta = 1.5\ \%$ and $S_2 = \text{III}$

functions respectively. In the same way, E_0, E_1 and E_2 are respectively the approximation errors of constant, linear and quadratic functions. One can note that column E_2 is the one that presents the smallest errors. However, in some cases, when the height is larger than 60 m the value of γ_{d2} can present non-realistic values. Because of that, we suggest to the readers and engineers to use the method

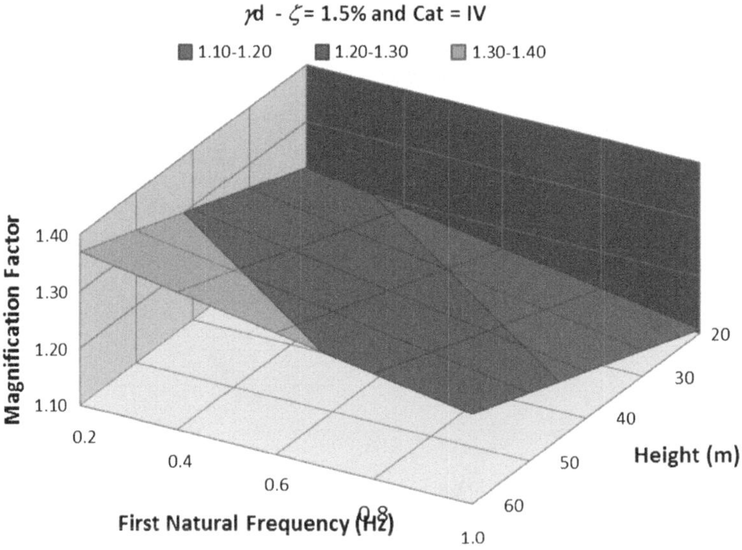

Fig. 9 Dynamic magnification factor γ_{d1} for $\zeta = 1.5$ % and $S_2 = \text{IV}$

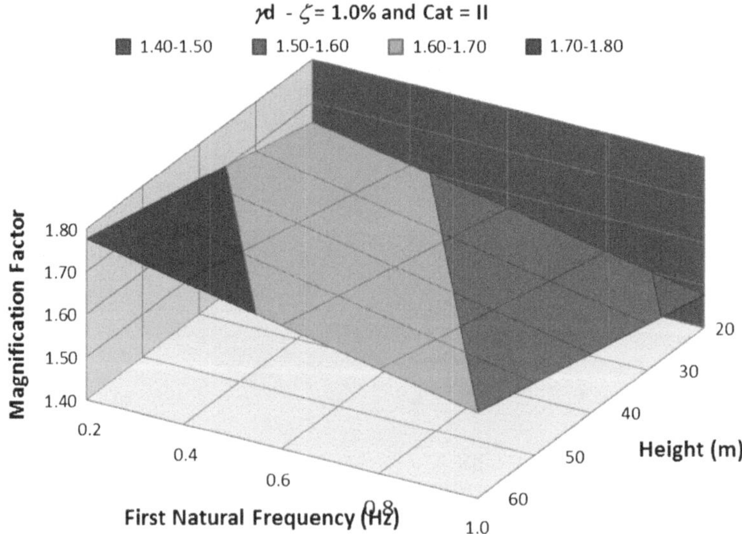

Fig. 10 Dynamic magnification factor γ_{d1} for $\zeta = 1.0$ % and $S_2 = \text{II}$

to adopt the linear approximation γ_{d1} and only consider the results presented here in structures no longer than 60 m. Tables 3, 4, 5, 6, 7, 8 show the coefficients of γ_{dk} for several values of ζ and S_2.

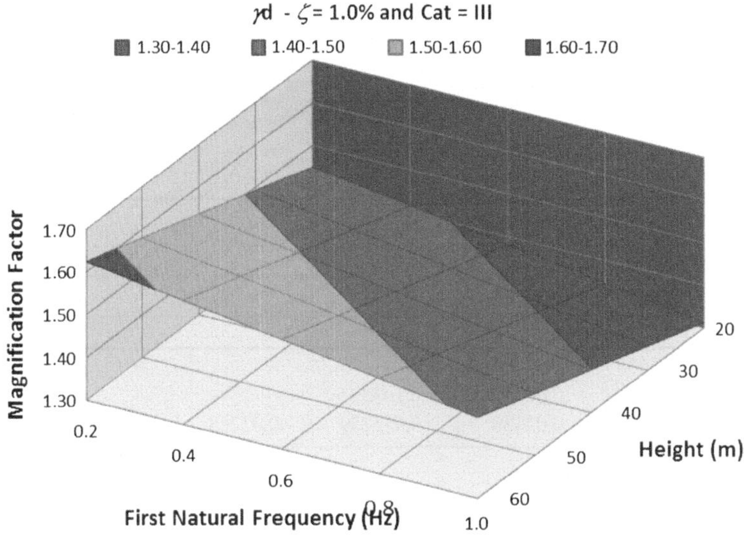

Fig. 11 Dynamic magnification factor γ_{d1} for $\zeta = 1.0$ % and $S_2 =$ III

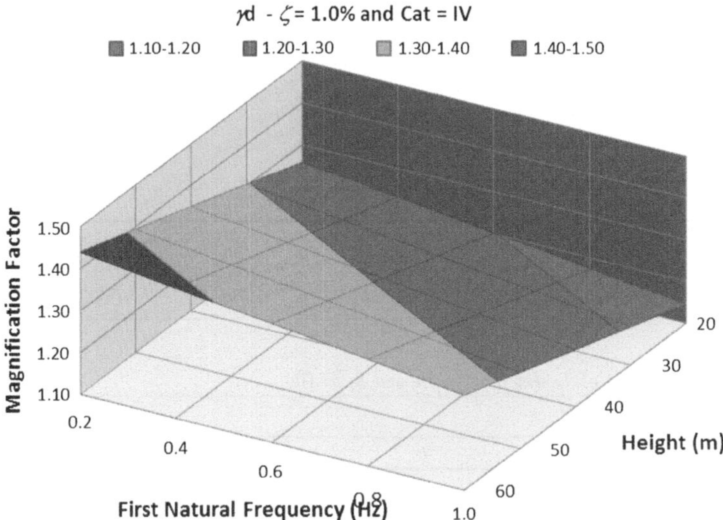

Fig. 12 Dynamic magnification factor γ_{d1} for $\zeta = 1.0$ % and $S_2 =$ IV

One can note that in the results shown in the present work, γ_{d} varies from 1.2 to 1.7. As we stated before, in this kind of structures, the main internal loads are the bending moment, the shear load and the axial load. Only the bending moment and

Table 3 Coefficients of γ_{dk} for $\zeta = 1.5$ % and $S_2 = \mathrm{II}$

	Design variables					
Function	*a*	*b*	*c*	*d*	*e*	*f*
0	1.596397	–	–	–	–	–
1	1.592341	0.002299	−0.20851	–	–	–
2	1.095788	0.026414	0.632866	−0.03026	−0.00021	−0.11634

Table 4 Coefficients of γ_{dk} for $\zeta = 1.5$ % and $S_2 = \mathrm{III}$

	Design variables					
Function	*a*	*b*	*c*	*d*	*e*	*f*
0	1.43633	–	–	–	–	–
1	1.3615	0.003594	−0.16339	–	–	–
2	0.943021	0.02439	0.575259	−0.02768	−0.00018	−0.08937

Table 5 Coefficients of γ_{dk} for $\zeta = 1.5$ % and $S_2 = \mathrm{IV}$

	Design variables					
Function	*a*	*b*	*c*	*d*	*e*	*f*
0	1.263054	–	–	–	–	–
1	1.172087	0.003751	−0.14004	–	–	–
2	0.79765	0.022412	0.50859	−0.02418	−0.00017	−0.08094

Table 6 Coefficients of γ_{dk} for $\zeta = 1.0$ % and $S_2 = \mathrm{II}$

	Design variables					
Function	*a*	*b*	*c*	*d*	*e*	*f*
0	1.67104	–	–	–	–	–
1	1.635324	0.003093	−0.22068	–	–	–
2	1.128192	0.029525	0.704224	−0.03643	−0.00024	−0.10375

Table 7 Coefficients of γ_{dk} for $\zeta = 1.0$ % and $S_2 = \mathrm{III}$

	Design variables					
Function	*a*	*b*	*c*	*d*	*e*	*f*
0	1.502536	–	–	–	–	–
1	1.393851	0.004403	−0.16899	–	–	–
2	0.94504	0.028501	0.636653	−0.03346	−0.00022	−0.06387

the shear load are multiplied by the magnification factor to obtain the dynamic results. In the cases analyzed here, the axial load is not influenced by the dynamic response.

Table 8 Coefficients of γ_{dk} for $\zeta = 1.0$ % and $S_2 = \text{IV}$

	Design variables					
Function	*a*	*b*	*c*	*d*	*e*	*f*
0	1.320372	–	–	–	–	–
1	1.1939	0.004552	−0.13956	–	–	–
2	0.795803	0.026117	0.571883	−0.02973	−0.0002	−0.0557

7 Conclusions

We presented linear static and dynamic models based on the NBR-6123 code [1] to compute the wind loads on telecommunication towers. A special emphasis was placed in pre-cast RC towers. A new procedure, based on graphs and curves obtained using optimization techniques, uses the results of the static analysis to compute the dynamic response of this kind of structures. One peculiar characteristic of these pre-cast RC structures is that they often present the first natural frequency of vibration smaller than 1 Hz and so the dynamic analysis is needed. The main feature researched is the dynamic magnification factor, defined here as the ratio between the bending moment given by the dynamic and static models. Surfaces are created to give the dynamic magnification factor as a function of the structure height and the first natural frequency of vibration. To create these surfaces, optimization problems (inverse problems) were formulated where the objective function is the error between the dynamic magnification factor and other given by equations, defined as a function of the structure height and first natural frequency of vibration. The design variables are the coefficients of these equations and constraints are imposed to avoid negative magnification factors. The methodology proposed here is quite precise and can reduce drastically engineering and computational time used to accomplish the dynamic analysis of RC Telecommunication Towers. Thus, the graphs created by the authors constitute a new tool that can easily be used by engineers.

As suggestions for futures works we propose the determination of a simplified model that takes into consideration the non-linearities of the structure.

Acknowledgments This work is the result of a research project completed during an academic visit accomplished by Marcelo Araujo da Silva at The University of Iowa, under the supervision of Prof. Jasbir S. Arora, in Iowa City, IA, USA. The program was sponsored by RM Soluções Engenharia Ltda. A great portion of the work was developed using resources of the Optimal Design Laboratory at The University of Iowa. All these supports are gratefully acknowledged.

References

1. ABNT—Associação Brasileira de Normas Técnicas. NBR-6123: Forças Devidas ao Vento em Edificações. Rio de Janeiro (1988) (in portuguese)
2. ABNT—Associação Brasileira de Normas Técnicas. NBR-6118: Projeto de Obras de Concreto. Rio de Janeiro (2003) (in portuguese)

3. ACI—American Concrete Institute. ACI Committee 318: Building Code Requirements for Reinforced Concrete. Detroit (1971)
4. Arora, J.S.: Introduction to Optimum Design, 2nd edn. Elsevier Academic Press, San Diego (2004)
5. Branson, D.E.: Instantaneous and time-dependent deflections of simple and continuous reinforced concrete beams. Alabama Highway Research Report, Bureau of Public Roads, Report No. **7**, Aug. 1963, pp. 1–78 (1963)
6. Brasil, R.M.L.R.F., Silva, M.A.: RC large displacements: optimization applied to experimental results. Comput. Struct. **84**, 1164–1171 (2006)
7. Chahande, A.I., Arora, J.S.: Optimization of large structures subjected to dynamic loads with the multiplier method. Int. J. Numer. Meth. Eng. **37**, 413–430 (1994)
8. Silva, M.A., Brasil, R.M.L.R.F.: Nonlinear dynamic analysis based on experimental data of RC telecommunication towers subjected to wind loading, Math. Prob. Eng. Article ID 46815 (2006)

Different Analysis Strategies for Roller Compacted Concrete Dam Design

Ugur Akpinar, Alper Aldemir and Baris Binici

Abstract This study presents the results of numerical simulations of the 56 m high Kocak concrete gravity dam planned to be constructed in Giresun, Turkey. A three stage analysis procedure was employed, namely rigid block stability analysis, two-dimensional finite element analysis both in frequency and time domain and three-dimensional dynamic analysis. The preliminary dimensions of the dam cross section were determined from stability analysis following recommendations in [1] and [2]. Afterwards, a time history analysis was conducted by following the procedure of [3] that accounts for foundation flexibility and dam-reservoir interaction using the finite element method. In order to estimate the crack lengths and distances, a nonlinear time history analysis was conducted for the two-dimensional model using Westergaard's [4] added mass approach including concrete cracking based on a smeared rotating crack approach. In the 3D linear spectrum analysis, the importance of modeling the exact geometry and complete soil-dam interaction and the influence of earthquake induced effects were investigated. It was found that the preliminary design section based on rigid body equilibrium is susceptible to cracking. However, the locations of expected cracks are highly dependent on the 2D versus 3D idealization of the dam geometry. Considering the fact that the length to height ratio of the dam is around 3, detailed three-dimensional simulations was found to be necessary to determine potential damage locations after an earthquake. The analysis results shed light on the locations of grout curtain to

U. Akpinar (✉) · A. Aldemir · B. Binici
Department of Civil Engineering, Middle East Technical University,
Dumlupınar Bulvarı, Ankara 06531, Turkey
e-mail: ugur.akpinar_02@metu.edu.tr

A. Aldemir
e-mail: aaldemir@metu.edu.tr

B. Binici
e-mail: binici@metu.edu.tr

A. Öchsner et al. (eds.), *Design and Analysis of Materials and Engineering Structures*, Advanced Structured Materials 32, DOI: 10.1007/978-3-642-32295-2_9,

reduce uplift pressures without sustaining severe damage under operational based earthquake hazard, and stability of the structure under maximum design and maximum credible earthquake hazard levels.

1 Introduction

This chapter presents, different analysis methodologies were utilized in the design process of roller-compacted concrete (RCC) dams emphasizing on a case study, i.e. Kocak dam, which is located in Camoluk region of Giresun and has a distance of nearly 9 km to the East Anatolian fault. The following analysis techniques were carried out throughout this study.

1. Rigid block stability analyses
2. 2D linear elastic frequency domain analyses
3. 2D nonlinear finite element analysis including dam-foundation-reservoir interaction
4. 3D linear elastic finite element analyses with multi directional earthquake effects including dam-reservoir-rock interaction

The aim of this study was to evaluate the safety of the most economical dam cross-section by comparing the stresses obtained from the aforementioned analyses with the design stress limits. The general approach followed in the course of the study is summarized in Fig. 1. The slopes of both downstream and upstream faces were taken as the main variables. Since the seismic effects play an important role, the local seismic risk is extremely important. In this context, site specific design spectrums and spectrum compatible synthetic ground motions generated by [5] were utilized. In light of seismic risk results, the factor of safety for different cross-section alternatives is determined by rigid block stability analyses. At this point, it should be mentioned that 2D and 3D finite element analyses are obviously more appropriate but more demanding for such preliminary analysis. For instance, stability analyses are suggested as only preliminary analyses technique by USACE [1, 2] recommendations.

After conducting the stability analyses, dynamic analyses were carried out by utilizing 2D and 3D finite element analysis (FEA). In the 2D analyses, hydrostatic and hydrodynamic forces, concrete crushing and dam-foundation interaction was also considered. In those detailed analyses, crack length of the dam base was limited so that it would not affect the drains of dam and grout curtain under the effect of operation based earthquake (OBE) whereas the dam should be in a repairable state after maximum design earthquake (MDE) hits the structure. Finally, stress checks were conducted for the three-dimensional dam-reservoir foundation interaction models by employing modal analysis. The stability of the 3D dam model is discussed in comparison to the 2D analyses results.

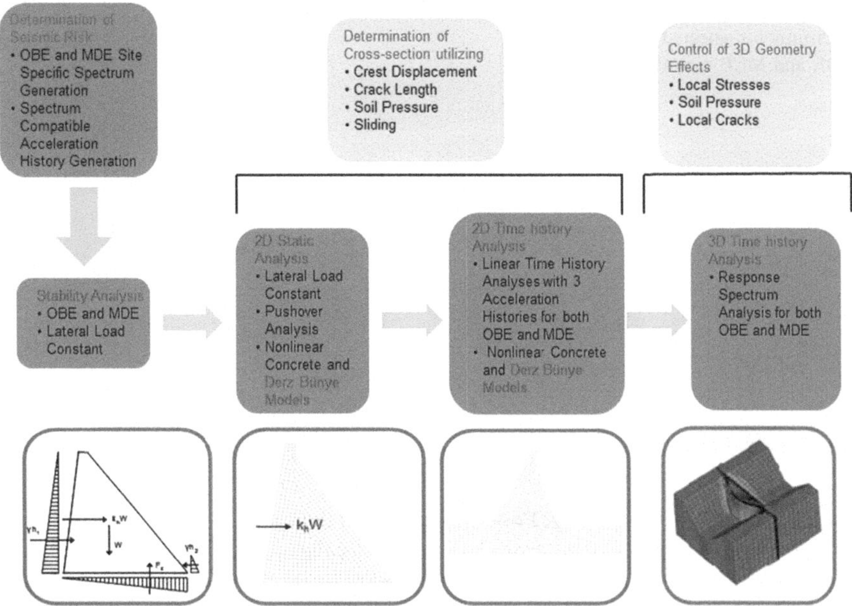

Fig. 1 Kocak dam seismic design approach

2 Analysis Parameters

2.1 Local Seismic Risk Results

Seismic performance is most probably the most important design aspect of the Kocak dam as it is located in a seismically active region (the first seismic zone defined by Turkish earthquake code 2007 [6]). Therefore, site specific earthquake demands should be determined. Figure 2 shows site specific spectrums for a rock soil profile having a shear-wave velocity of 760 m/s. In Turkish earthquake code 2007 [6], peak ground acceleration and spectral acceleration for the first seismic zone, are suggested as 0.4 and 1.0 g, respectively. It is clear from Fig. 2 that the site specific spectral values for OBE, MDE and maximum credible earthquake (MCE) are nearly 80, 125 and 200 % of the Turkish earthquake design spectrum values [6].

The site specific spectrums cannot be utilized when performing time history analyses. Thus, spectrum compatible acceleration time histories were generated [5]. These six synthetic ground motions for the most critical governing seismic events are presented in Fig. 3.

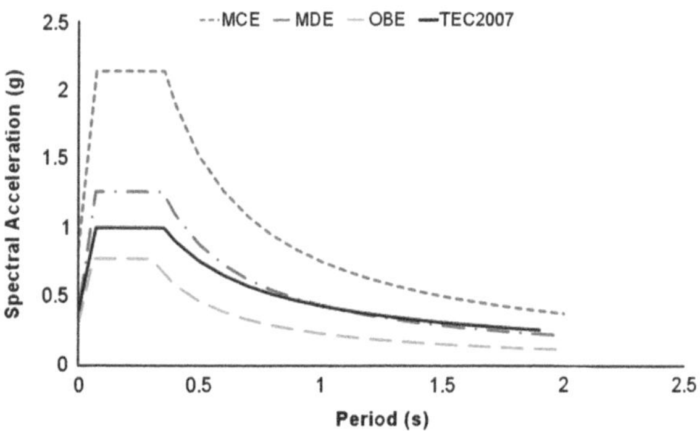

Fig. 2 Site specific spectrums for selected OBE, MDE and MCE records

2.2 Material Properties

Average modulus of elasticity, Poisson's ratio and compressive strength values for the foundation rock obtained from material tests carried out at the site are presented in Table 1. The alluvium fill that observed in site is a granular material and its modulus of elasticity, unit weight and Poisson's ratio are assumed to be 200 MPa, 18 kN/m^3 and 0.3.

C20 class concrete was used for the dam body. The allowable tensile strength for this class is calculated as 2.1 MPa for OBE and 3.1 MPa for MDE/MCE earthquakes by utilizing [1, 2] and methods described in [7]. As these studies give an approximate value of tensile strength due to the implementation of empirical formulas, the results were checked with split tensile tests conducted on the concrete mixture similar to the one used in the site. The test results show that the tested specimens' tensile strength value (1.62 MPa) is compatible with the ones calculated from the literature formulas. In the light of the above data, dynamic tensile limits are 2.1 MPa for OBE and 3.10 MPa for MDE/MCE earthquakes. The modulus of elasticity values for concrete was calculated from [1] and [8] resulting in 23.75 GPa. The unit weight of concrete was taken as 24 kN/m^3. Although the dam rests on rock, the modulus of elasticity of the foundation is only half that of the dam body. Therefore, the foundation flexibility may result in period elongation and radiation damping as noted by [9].

In document [1], which is mainly based on the studies carried out by Chopra and Fenves [3], the damping ratio for FEM analysis with massless foundation is recommended to be computed from three factors affected by the structure, foundation and reservoir. From this study, the damping ratio was calculated as 10 %. For the time history analyses, the Rayleigh damping was adjusted so that it gives 10 % damping ratio between the first and third mode of vibrations.

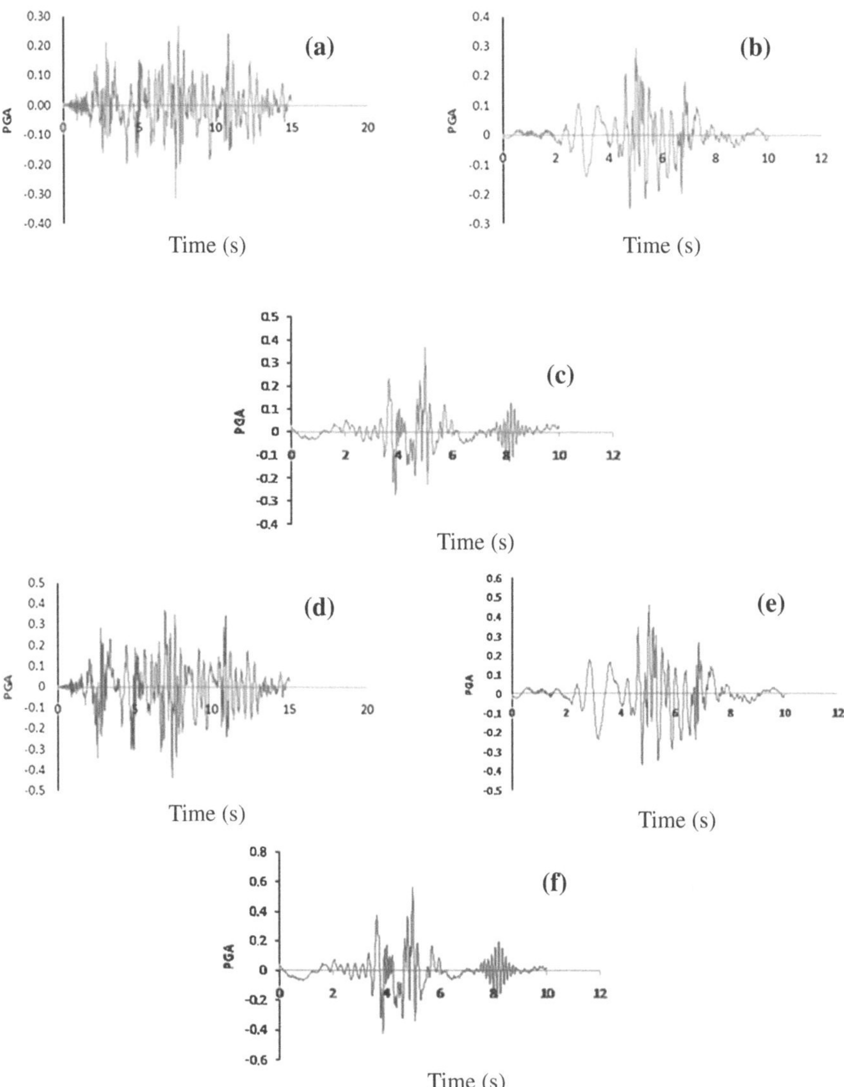

Fig. 3 OBE (**a–c**) and MDE (**d–f**) acceleration histories (units: g)

Table 1 Average modulus of elasticity and average Poisson's ratio values

	Modulus of elasticity (GPa)	Unit weight (kN/m^3)	Poisson's ratio	Compressive strength (MPa)
Average	11.80	26.24	0.22	40.30
Standard deviation	3.05	0.32	0.02	21.73

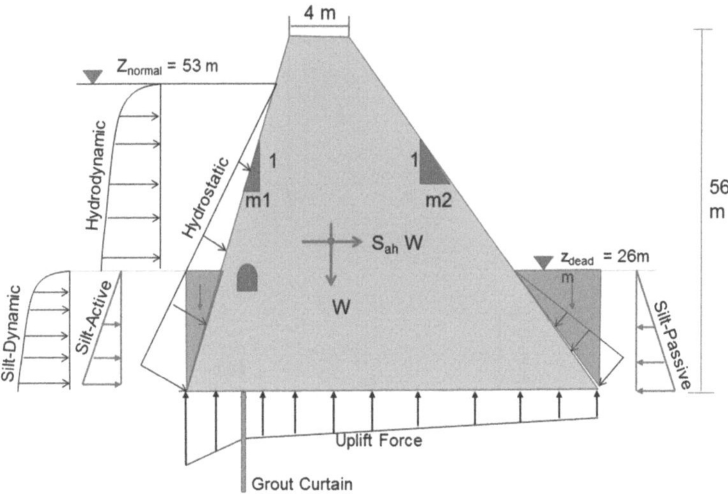

Fig. 4 Forces in stability analysis of a generic dam section

3 Design Models

3.1 Stability Analysis

Stability analysis is an effective tool for the preliminary design of dams. The analysis was applied following the recommendations in [2]. This procedure assumed the dam's cross-section as a rigid block and tests the model for slipping and overturning. Details of the dam's cross-section and applied forces are shown in Fig. 4.

Figure 4 shows the applied forces on the rigid block namely the self-weight of the dam, the lateral earthquake load, the hydrostatic and hydrodynamic forces, the active and passive pressures and the dead load of the silt and uplift forces. The hydrodynamic forces were calculated according to Westergaard's [4] added mass approach by the equation:

$$p_Z = \frac{7}{8} \rho \ddot{u}_g \sqrt{H(H - Z)} \tag{1}$$

For the calculation of the uplift pressure, the effect of the grout curtain was taken into account by reducing the effect pressure at the drain level by a factor of 2/3. Lateral earthquake loads were determined for OBE (0.6 g) and MDE (0.97 g) by employing the 10 % damped design response spectrum (Fig. 2). The fundamental period of the dam was taken from the 3D analysis as 7 Hz. For the sliding checks, a cohesive strength was taken as 0.9 MPa by matching the direction tension and compression test results to a Mohr–Coulomb failure envelope resulting in a friction angle of 1.5 MPa and 0.6, respectively. Stability analyses were conducted for the four different loading cases:

Table 2 Factor of safety values for overturning and sliding

Case	$FS._{slide}$ (ratio of slipping resistance to slipping effect)	$FS._{overturning}$ (B: distance between resultant force and center of gravity)
Normal loading	2.0	B/6
Irregular loading	1.7	B/4
Extreme loading	1.3	B/2

Table 3 Allowable stresses of concrete for different loading cases

Case	Compression	Tension	Unit
Normal loading	$0.3 \times f_c = 6.0$	0	MPa
Irregular loading	$0.5 \times f_c = 10.0$	$0.6 \times f_t = 1.2$	MPa
Extreme loading	$0.9 \times f_c = 18.0$	$1.5 \times f_t = 3.0$	MPa

1. End of construction (empty reservoir) + OBE
2. Operating (full reservoir) + OBE
3. Operating (full reservoir) + MDE
4. Flood with 56 m reservoir level

Factor of safeties for overturning and slipping and for allowable stresses on concrete dams are given in Tables 2 and 3, respectively.

3.2 Finite Element Analysis

Finite element analyses were conducted for the following three cases:

1. Linear analysis in frequency domain by employing vertical and horizontal earthquake loads and hydrodynamic loads on a rigid foundation dam [9].
2. Nonlinear history analysis in time domain by employing the added mass approach for hydrodynamic forces, flexible half-space foundation and concrete cracking.
3. Linear spectrum analysis for the 3D model employing the added mass approach for hydrodynamic forces with flexible massless foundation.

All the modeling approaches are summarized in Table 4. Their schematic representations are also given in Fig. 5. As can be seen from the table, the non-linear time history analysis gives the most comprehensive results whereas the stability analysis provides very limited information. To have an accurate simulation of the dam, there are two important phenomena that should be included in the model. The first one is the movement of the concrete dam on the foundation. The second one is modeling the cracks that may occur in the concrete dam's body. In the 2D nonlinear time history analysis, cracking in concrete and its direction could be modeled by employing nonlinear material models. A mathematical model that covers all the parameters in a 3D nonlinear finite element model on flexible foundation with transmitting boundaries is not practical due to immense need of

Table 4 Comparison of different modeling methodologies

Analysis type	Dam's deformation	Ground deformation	Earthquake effect	Size effect	Concrete cracking effect	Silt effect	Hydrodynamic effect
Stability	X	X	X	X	X	√	~
Analysis method 1	√	X	√	X	X	X	√
Analysis method 2	√	√	√	X	√	√	~
Analysis method 3	√	√	√	√	X	X	~

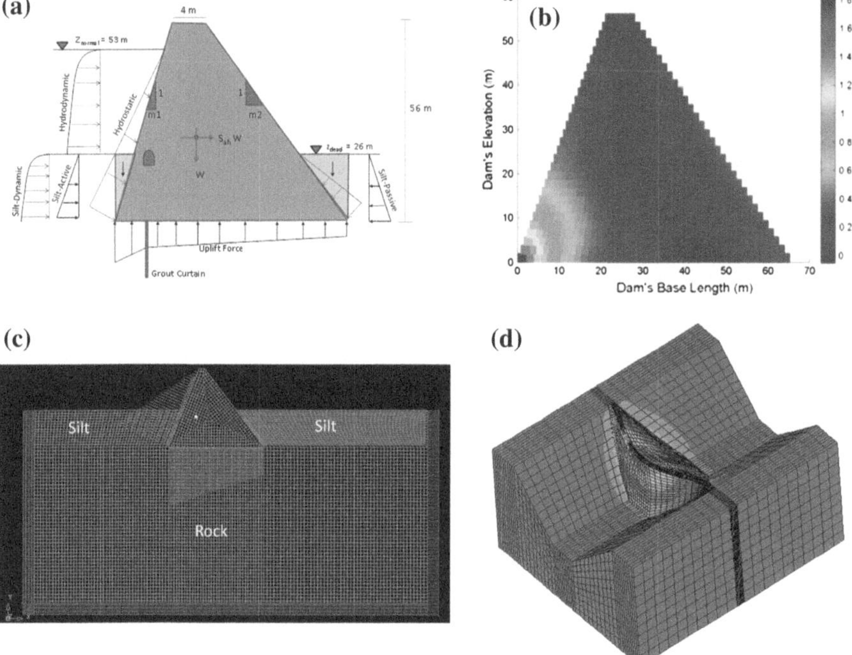

Fig. 5 Illustration of the analysis models. **a** Stability model. **b** Frequency domain solution. **c** Nonlinear time history analysis. **d** 3D Spectrum analysis

computational power. Instead, the results of each model are examined separately and employed for the design of the dam section.

In order to simulate the nonlinear concrete behavior, a total strain rotating crack model as explained in detail by [10] was applied. Compressive behavior is defined by a parabolic stress–strain relationship. The crushing behavior and ultimate strain

Table 5 Stability analysis for slopes $m_1 = 0.7$, $m_2 = 0.7$

	Overturning				Slipping		
	e_{safety}	e^*	f_{safety}		f_{safety}	f_{slip}	
End of construction							
Const. End	20.8	–	–	OK	1.7	–	OK
+ EQ (OBE)	20.8	12.29	1.69	OK	1.7	3.39	OK
+ EQ (OBE)	20.8	12.29	1.69	OK	1.3	3.39	OK
Operating							
Operating	13.9	0.72	19.31	OK	2	21.59	OK
+ EQ (OBE)	20.8	15.58	1.34	OK	1.7	3.13	OK
+ EQ (MDE)	41.7	24.26	1.72	OK	1.3	1.94	OK
Flood							
Flood	41.7	0.44	94.77	OK	1.3	55.05	OK

e^* eccentricity at the base of the dam

Table 6 Stability analysis for slopes $m_1 = 0.4$, $m_2 = 0.8$

	Overturning				Slipping		
	e_{safety}	e^*	f_{safety}		f_{safety}	f_{slip}	
End of construction							
Const. end	17.9	–	–	OK	1.7	–	OK
+ EQ (OBE)	17.9	8.32	2.15	OK	1.7	3.39	OK
+ EQ (OBE)	17.9	16.26	1.10	OK	1.3	3.39	OK
Operating							
Operating	11.9	2.2	5.41	OK	2	17.44	OK
+ EQ (OBE)	17.9	15.04	1.19	OK	1.7	3.10	OK
+ EQ (MDE)	35.7	24.21	1.47	OK	1.3	1.91	OK
Flood							
Flood	35.7	0.34	105.00	OK	1.30	46.25	OK

are governed by the compressive fracture energy so that the ultimate crushing strain in the models is meshing independent. The tensile behavior was modeled by using a linear softening function beyond the tensile strength.

4 Analysis Results

4.1 Stability Analysis

The results of the stability analysis using the method described above are given in Tables 5, 6, 7, 8. First the analysis was conducted with a dam section with $m_1 = 0.7$ and $m_2 = 0.7$ where m_1 is the upstream slope and m_2 is the downstream slope of the dam. According to the results, this section has approximately a safety factor of 1.34 in the most critical case. Reducing the upstream slope (m_1) to 0.4

Table 7 Stability analysis for slopes $m_1 = 0.4$, $m_2 = 0.7$

	Overturning				Slipping		
	e_{safety}	e^*	f_{safety}		f_{safety}	f_{slip}	
End of construction							
Const. end	16.4	–	–	OK	1.70	–	OK
+ EQ (OBE)	16.4	9.32	1.76	OK	1.70	3.39	OK
+ EQ (OBE)	16.4	15.27	1.07	OK	1.30	3.39	OK
Operating							
Operating	10.9	0.99	11.01	OK	2.00	16.67	OK
+ EQ (OBE)	16.4	16.85	0.97	Not OK	1.70	3.04	OK
+ EQ (MDE)	32.7	26.35	1.24	OK	1.30	1.87	OK
Flood							
Flood	32.7	0.95	34.42	OK	1.30	42.30	OK

Table 8 Stability analysis for slopes $m_1 = 0.3$, $m_2 = 0.7$

	Overturning				Slipping		
	e_{safety}	e^*	f_{safety}		f_{safety}	f_{slip}	
End of construction							
Const. end	16.4	–		OK	1.70	–	OK
+ EQ (OBE)	16.4	7.34	2.23	OK	1.70	3.39	OK
+ EQ (OBE)	16.4	17.25	0.95	OK	1.30	3.39	OK
Operating							
Operating	10.9	1.75	6.23	OK	2.00	16.51	OK
+ EQ (OBE)	16.4	16.59	0.99	Acceptable	1.70	3.01	OK
+ EQ (MDE)	32.7	26.35	1.24	OK	1.30	1.85	OK
Flood							
Flood	32.7	0.38	86.12	OK	1.30	41.75	OK

and increasing the downstream slope (m_2) to 0.8 decreases factor of safety to 1.1. In the next step, m_2 slope was reduced to 0.7 while keeping the m1 slope the same. For these slopes, it is observed that the factor of safety of this section is slightly below 1.0 for the overturning case. For all cases, it was assumed that a 26 m high silt layer is placed to reservoir after the construction of the dam is completed. However, placing the silt layer during the construction may increase the stability. With this assumption, a dam section with $m_1 = 0.3$ and $m_2 = 0.8$ was investigated. According to stability analysis results, this section satisfies the factor of safety limits presented in Table 2.

4.2 Linear Finite Element Analysis in the Frequency Domain

The frequency domain analyses were conducted by employing EAGD84 [11] which is a computer program for earthquake analysis of concrete gravity dams. For all the ground motion records, tip displacement of the dam, maximum tension

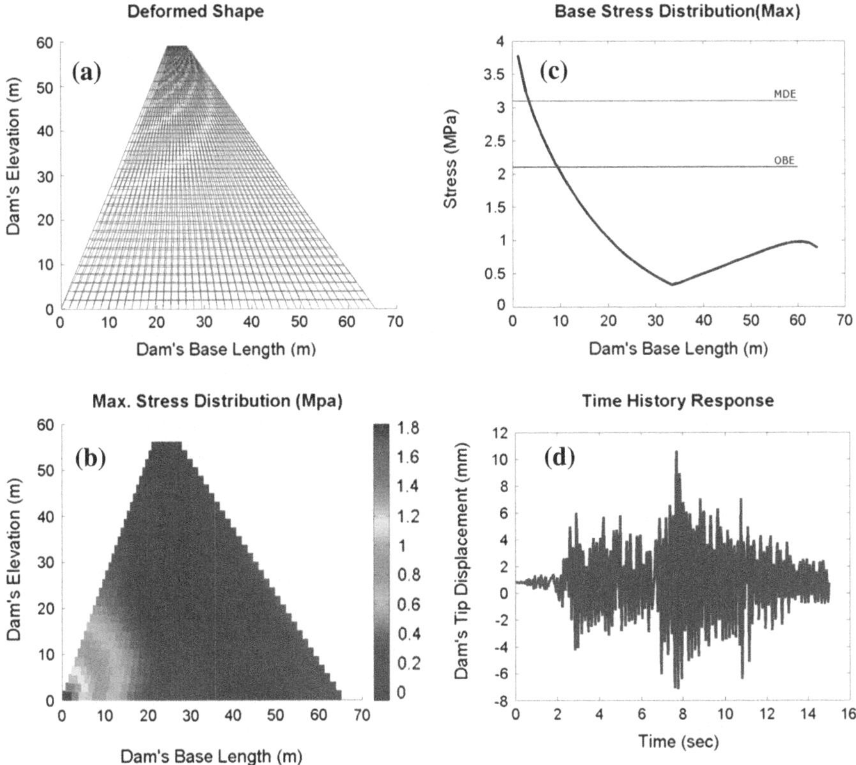

Fig. 6 Results of frequency domain (2D) analysis. **a** Deformed shape. **b** Stress distribution on dam. **c** Stress distribution on the base of the dam. **c** Tip displacement history

stress distribution on the dam, stress distribution on the base of the dam and corresponding limit states were computed (Fig. 6). Analyses were also repeated with an additional vertical ground motion. For these cases, 80 % scale of same ground motion (horizontal) was utilized as vertical ground motion. The results obtained for the different dam sections are given in Table 9 along with the percent of limit stress exceedance levels.

According to the analysis results of all sections, the maximum tensile stress is below the limit under the effect of only horizontal OBE ground motions. If the vertical component is considered, displacements and stresses may increase rapidly. The highest level of tensile stress is found to be about 25 % higher than the dynamic tensile strength of 20 MPa of concrete at the dam heel. This suggests conducting nonlinear time history analysis and use a higher strength concrete at the dam heel if needed.

Table 9 Effects of dam's slopes on stress and tip displacement results

		Δ (m)	σ_{max} (MPa)	σ_{min} (MPa)	Δ (m)	σ_{max} (MPa)	σ_{min} (MPa)
			$m_1 = 0.4, m_2 = 0.7$			$m_1 = 0.4, m_2 = 0.8$	
Ground Motion	OBE1	5.444	1.822	−1.76	4.267	1.526	−1.58
	OBE2	3.624	1.313	−1.75	2.974	1.129	−1.68
	OBE3	5.069	1.734	−1.73	4.746	1.671	−1.66
	MDE1	10.59	3.785	−3.01	7.986	3.288	−2.78
	MDE2	5.953	1.907	−2.52	5.307	1.671	−2.39
	MDE3	7.934	2.694	−2.53	7.376	2.603	−2.38
Combo	OBE1	7.855	2.528	−2.11	6.849	2.295	−2.09
	MDE1	11.81	3.651	−2.72	9.266	3.026	−2.82
			$m_1 = 0.35, m_2 = 0.8$			$m_1 = 0.4, m_2 = 0.8$	
Ground Motion	OBE1	5.148	1.864	−1.69	4.267	1.526	−1.58
	OBE2	3.368	1.308	−1.79	2.974	1.129	−1.68
	OBE3	4.907	1.798	−1.77	4.746	1.671	−1.66
	MDE1	9.873	3.916	−3.16	7.986	3.288	−2.78
	MDE2	5.849	1.925	−2.62	5.307	1.671	−2.39
	MDE3	7.711	2.825	−2.62	7.376	2.603	−2.38
Combo	OBE1	7.675	2.632	−2.11	6.849	2.295	−2.09
	MDE1	11.38	3.783	−2.83	9.266	3.026	−2.82

OBE: Allowable tension stress (dynamic): 2.1 Mpa
MDE: Allowable tension stress (dynamic): 3.1 Mpa

4.3 Nonlinear Analysis

The linear analysis provides no information on the expected damage. For a detailed investigation, nonlinear time history analysis should be conducted in order to find crack locations widths and extends. For the given ground motion records and dam sections, tip displacement of the dam, location of cracks on the dam, total crack width and length were determined (Fig. 7). The analysis results of the dams with varying upstream and downstream slopes are given in Table 10. As can be seen from results, the design earthquakes, OBE and MDE cause around 3–5 and 9–13 m crack lengths, respectively, at the dam base. The maximum crack widths are expected to remain between 1.2 and 2.5 mm. The nonlinear analysis suggests that crack width and crack length are mostly affected from the ground motion related parameters, and the upstream and downstream slopes of the dam do not significantly influence the expected extent of cracking.

4.4 3D Finite Element Analyses

The 3D representation of the dam model contains approximately 80,000 elements to determine the local stress concentrations. Since 0.3–0.8 slopes satisfy the design limits for analyses performed up to this point, this section is utilized in the 3D

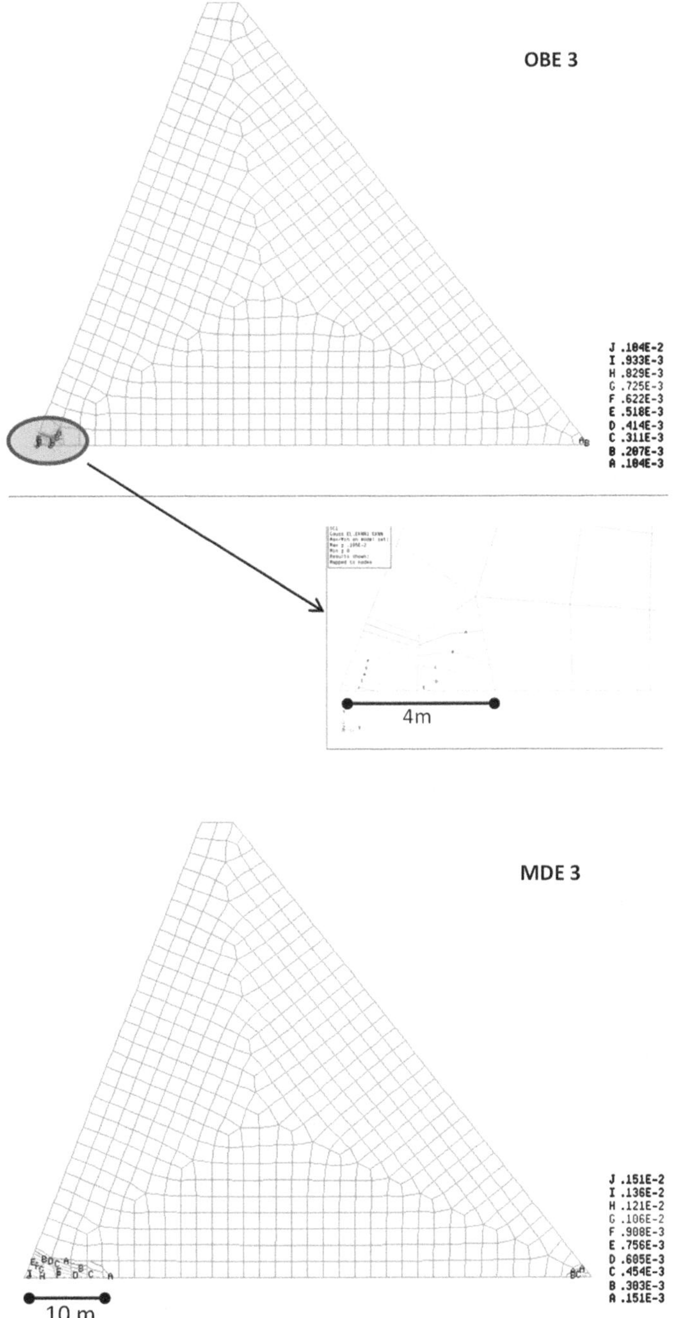

Fig. 7 Propagation of cracks on 2D analytical model of selected dam section

Table 10 Crack lengths on concrete dam section

		Tip disp. (mm)	Max. stress (MPa)	Crack length (m)	Crack with (mm)
$m_1 = 0.4, m_2 = 0.7$					
Ground motion	OBE1	12.6	0.96	3.98	1.15
	OBE2	12.7	0.93	3.98	1.14
	OBE3	14.3	0.94	4.97	1.31
	Average	13.2	0.94	4.31	1.21
	MDE1	18.9	0.96	9.94	2.14
	MDE2	19.6	0.91	10.9	1.95
	MDE3	21.8	0.92	12.9	2.01
	Average	20.1	0.93	11.3	2.03
$m_1 = 0.3, m_2 = 0.8$					
Ground motion	OBE1	12.4	0.97	3.98	1.18
	OBE2	12.7	0.93	3.98	1.17
	OBE3	14.3	0.95	4.97	1.36
	Average	13.1	0.95	4.31	1.24
	MDE1	19.1	0.92	9.94	2.23
	MDE2	19.6	0.88	10.93	2.17
	MDE3	22.1	0.88	12.92	1.83
	Average	20.2	0.89	11.93	2.08
$m_1 = 0.4, m_2 = 0.8$					
Ground motion	OBE1	11.3	0.96	3.98	1.12
	OBE2	11.3	0.91	3.98	1.08
	OBE3	12.8	0.93	4.97	1.24
	Average	11.8	0.93	4.31	1.14
	MDE1	17.2	0.94	8.9	2.06
	MDE2	17.7	0.92	10.88	1.99
	MDE3	19.8	0.91	12.86	1.71
	Average	18.2	0.92	10.88	1.92

analyses. For the sake of less computational power demand, the response spectrum analysis method was selected to determine the earthquake effects. First, the natural vibration periods and shapes of the 3D model were found. The results for the first three modes are shown in Fig. 8. It is clear from this figure that the natural frequencies of the modes are very close to each other, hence requiring the consideration of many modes with appropriate combination rules. The modal contributions were combined by utilizing the complete quadratic combination (CQC) rule. In these analyses, three directional earthquake effects were considered along with the vertical earthquake effects. The earthquake loading combinations were formed by using the following rule. The lateral load factors were taken as 0.30 and 1 for dam axis direction (x-axis) and stream directions (z-axis), varyingly. Moreover, the load factor for vertical effects was assumed to be 0.80 as this dam is very close to the fault zone and its natural vibration periods are very low.

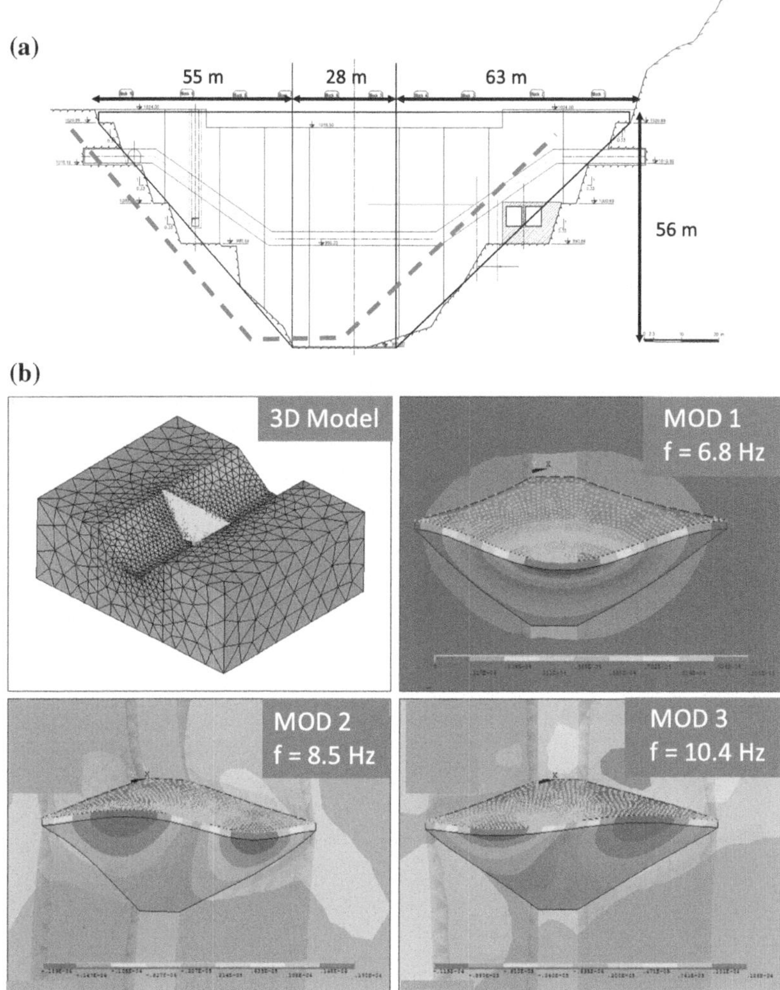

Fig. 8 3D finite element model of selected dam section and its natural mode shapes. **a** Dam axis plan and idealization for model. **b** 3D model and vibration periods

The maximum and minimum principal stress results obtained from the response spectrum analyses are summarized in Table 11 and Figs. 9, 10. It can be inferred from Table 3 that the tensile limit is barely exceeded. This situation does not hold for MDE and MCE. For MDE earthquake, the principal stresses are larger than the tensile limits but the damage is not expected to be significant as the overstressed regions do not spread all over the dam body (Fig. 10a, b). Furthermore, the estimated cracked region is observed at about 10 m if the regions exceeding the tensile stress limits are assumed to be the crack length.

Table 11 Response spectrum analysis results of 3D analytical model

	x	z	x + 0.3z + 0.8y	0.3x + z + 0.8y
OBE				
u_x (mm)	3.17	1.31	3.16	1.65
u_y (mm)	−2.57	−2.34	−1.31	−1.35
u_z (mm)	2.25	9.01	3.87	9.35
σ_1 (MPa)	1.82	2.05	1.82	2.23
σ_3 (MPa)	−1.31	−1.01	−0.98	−1.01
MDE				
u_x (mm)	4.99	1.80	4.99	2.29
u_y (mm)	−2.55	−2.19	−0.73	1.22
u_z (mm)	3.08	13.91	5.57	14.48
σ_1 (MPa)	2.61	3.45	2.72	3.76
σ_3 (MPa)	−1.66	−1.06	−1.07	−1.06
MCE				
u_x (mm)	8.37	2.72	8.38	3.54
u_y (mm)	−2.53	2.64	1.80	3.78
u_z (mm)	4.59	22.78	8.67	23.76
σ_1 (MPa)	4.34	6.03	4.67	6.56
σ_3 (MPa)	−2.35	−1.17	−1.38	−1.16

u_x : Tip displacement in x-direction
u_y : Tip displacement in y-direction
u_z : Tip displacement in z-direction
σ_1 : Principle stress in 1-direction
σ_3 : Principle stress in 3-direction

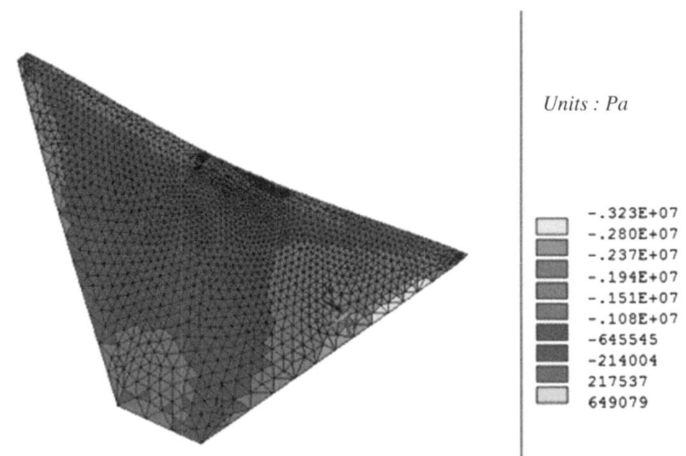

Fig. 9 Maximum compressive stress on selected dam section for MDE

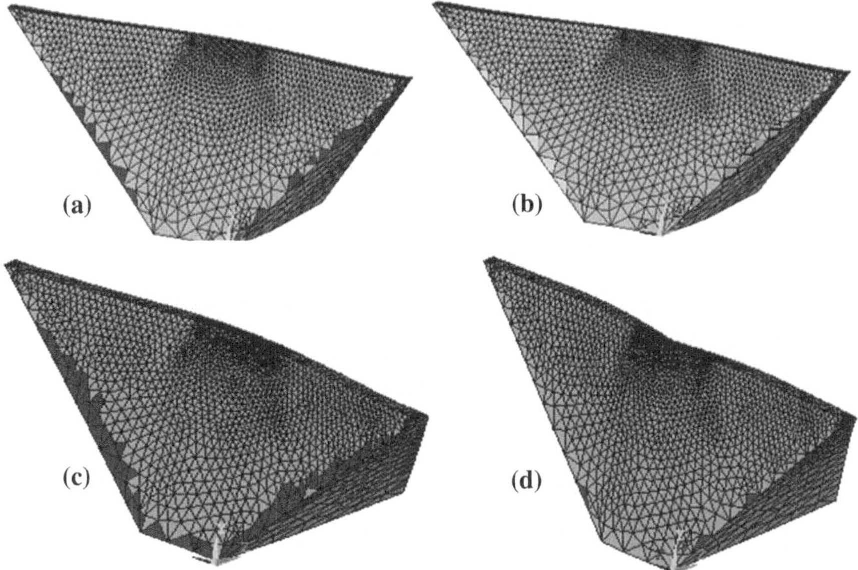

Fig. 10 Principal stresses on analyzed dam that greater than tensile limits (3.1 MPa) for MCE and MDE. **a** MDE 0.3X + 0.8Y + Z combination. **b** MDE-Z combination. **c** MCE 0.3X + 0.8Y + Z combination. **d** MCE-Z combination

5 Conclusions

The results of this four stage analyses study are summarized as follows:

1. A upstream slope of 0.3 and a downstream slope of 0.8 would satisfy the minimum damage limits for OBE and collapse prevention limits for MDE.
2. The maximum crest displacements for OBE and MDE were estimated as approximately 10 and 15 mm, respectively. Moreover, these earthquakes would cause a principal compressive stress of 2.5 and 3.5 MPa on the soil, respectively.
3. In 2D analyses, the crack width at the base was calculated as 4 m for OBE and 12 m for MDE. These values were estimated as at most 10 m in 3D analyses. Hence, the expected crack length obtained from 2D nonlinear analysis and 3D linear analysis are similar. Such a cracking is not expected to result in any damage for the drains.
4. The multi directional application of earthquake results in the worst condition, as far as the dam body stresses are concerned. Therefore, the seismic risk is controlled under the effect of multi directional earthquake loading.
5. The three-dimensional models give the most reliable results as the width-to-height ratio of Kocak dam is approximately 3. For instance, the most critical sections appear to be the right and left slopes of the dam, which could not be observed from the two dimensional analyses.

6. In order to limit the tensile stresses and cracks, it is recommended that a higher class concrete, i.e. C25, can be utilized at the base of this dam (up to approximately 9 m), while a lower strength concrete may be utilized in the upper dam body.

References

1. United States Army Corps of Engineers (1995) Seismic design provisions for roller compacted concrete dams. Engineering procedure 1110-2-12
2. United States Army Corps of Engineers (1995) Seismic design provisions for roller compacted concrete dams. Engineering procedure 110-2-2200
3. Fenves, G., Chopra, A.K.: Simplified analysis for earthquake resistant design of concrete gravity dams. Earthquake Engineering Research Center, University of California, Berkeley Report No. UCB/EERC-85/10 (1985)
4. Westergaard, H.M.: Water pressures on dams during earthquakes. Trans. Am. Soc. Civil. Eng. **98**, 418–433 (1933)
5. Yılmaz, T.: Koçak barajı için tasarım spektrumunun olasılık hesaplarına dayalı simik tehlike analizi (2011)
6. Ministry of Public Works and Settlement Government of Republic of Turkey: Specification for structures to be built in disaster area. Ankara, Turkey (2007)
7. Harris, D.W., Mohorovic, C.E., Dolen, T.: Dynamic properties of mass concrete obtained from dam cores. ACI Mater. J. **97**(3), 290–296 (2000)
8. American Concrete Institute: Building code requirements for structural concrete. ACI 318-05, ACI Committee 318, Detroit (2005)
9. Fenves, G., Chopra, A.K.: EAGD84: a computer program for analysis of concrete gravity dams. Report No. UCB-EERC/84-11 (1984)
10. Vecchio, F.J., Collins, M.P.: The modified compression field theory for reinforced concrete elements subjected to shear. ACI J. **83**(2), 219–231 (1986)
11. Fenves, G., Chopra, A.K.: EAGD-84: a computer program for earthquake analysis of concrete gravity dams. UCB/EERC-84/11, Earthquake Engineering Research Center, University of California, Berkeley (1982)

Experimental Method for Explosion Effect Determination

Jonathan Camargo and Luis Ernesto Muñoz

Abstract The study of the fragmentation originated from explosions is a challenging task, considering the conditions in which the phenomena occur. Those conditions are directly related with the nature of the explosion, which generates a high speed response of every part of the system; including dynamic behaviours from the chemical, mechanical, and aerodynamical point of view. This study presents an experimental approach to the determination of fragmentation characteristics, isolating the fragmentation effects from the shockwave. Based on standard *ITOP 4-2-813,* measurement methodology and instrumentation device were developed and implemented. This standard provides simple guidelines for designing experiments for explosion effects, taking into account the symmetric geometry of the explosive specimen for simplifying data recollection, by measuring mass-size in one half of a test arena and velocity of fragments in the opposite symmetric half. Velocity was assessed by microcontroller driven electronic hardware for which a custom barrier sensor was designed for manufacturing with single layer thin (thickness <0.3 mm) FR-4 copper clad. The speed reduction of a typical fragment was verified by simulation using coupled SPH-Lagrange. Finally, a sample experiment was done for checking the operation of the system, finding an ease of use in field.

J. Camargo (✉) · L. E. Muñoz
Department of Mechanical Engineering, Universidad de los Andes, Bogotá, Colombia
e-mail: jon-cama@uniandes.edu.co

L. E. Muñoz
e-mail: lui-muno@uniandes.edu.co

A. Öchsner et al. (eds.), *Design and Analysis of Materials and Engineering Structures*, Advanced Structured Materials 32, DOI: 10.1007/978-3-642-32295-2_10,

1 Introduction

Due to the complexity and variety of shapes and all the different locations of a target in a blast, it is necessary to estimate the probability of damage in an indirect manner by measuring the dynamic variables that describe the phenomenon in some significant points and extrapolating to the whole area of analysis. Using data that relate those variables with the damage associated to a target, one can predict the destructive capacity in a specified detonation condition.

For multiple purpose bombs, the detonation can be divided into two different phenomena:

Fragmentation: after detonation, the bomb casing is divided into fragments that travel at different speeds in a range from hundreds to thousands meters per second.
Shock wave: the supersonic combustion process that happens in the detonation produces the propagation of a high speed wave which transports an important part of the energy of the chemical reaction.

Different studies focused on explosion shockwave measurement are found in literature [1–3]. Few have considered fragment behaviour but aiming to the wound effects via simulation and medical testing [4–6]. This study presents an experimental approach to the determination of fragmentation characteristics, isolating the fragmentation effects from the shockwave by measuring the fragments' features: velocity, mass, and distribution in space. A method to determine speed of fragments is established, for which a timer circuit and a disposable sensor are developed with the aim of producing reliable but low cost instrumentation. This would allow executing tests on improvised explosive devices which are highly dangerous for civil population, and require an increasing number of experiment reproductions.

2 Methodology

In order to simplify the measuring process, the trajectories are considered parallel to the floor (i.e. fragments in a 2D space) and radial to the detonation locus, which reduces the aim of experiment to obtain the speed and mass in different locations of the plain field. The sketch in Fig. 1 presents the important data to be assessed.

In the center of Fig. 1, the bomb is presented in dotted blue, and at different sites a perimeter is drawn that corresponds to the points of observation of fragments in which the mass and velocity have to be measured. One should note that even if the radius is constant in Fig. 1, it is not mandatory to place measuring points in a circular perimeter, although it simplifies data processing and execution of a test.

To choose an effective way of measuring, the testing procedure protocol *ITOP 4-2-813* [7] was employed. This protocol was developed by the USA, the UK and Germany to provide a general procedure for determining velocities, masses and distribution of fragments in static detonation.

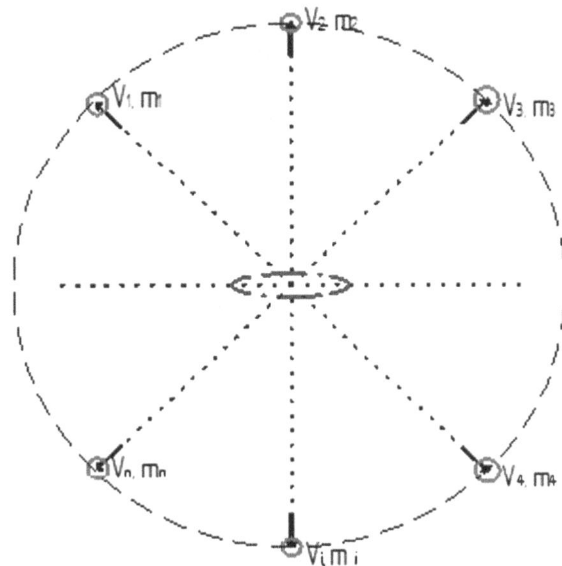

Fig. 1 Sketch of the travelling fragments phenomenon (*upper view*). To completely characterize the fragmentation, mass and velocity of fragments have to be measured in every spot of the field

The field area is selected according to the destructive capacity of detonation, related to the mass of explosives. This area is commonly known as the *fragmentation arena* and is the zone in which different measuring systems must be placed. A further simplification to the perimeter is made considering the explosion as symmetric with respect to the axis of the bomb; it can be assumed that what happens on one side of the symmetry axis is equal on the opposite side. This simplification leads to the reduction of the number of devices employed by measuring one set of properties in each of the sides, as shown in Fig. 1.

In the sketch in Fig. 2 one can see what this consideration implies: the horizontal line divides the field in two equal parts, in the upper side velocity data is collected and in the opposite side mass and distribution. To define the size of the arena, the data in Table 1 is suggested by the *ITOP 4-2-813* protocol [7]:

Mass and distribution are easily measured by observation, counting and weighting of fragments in the recuperation zone. This perimeter can be a simple wall structure made of wood or other materials that can hold fragments after the explosion, thus the size and thickness of the planks that make the walls depend on the mass of explosive and approximate size of the fragments. A complete virtual scenario of testing is presented in Fig. 3.

2.1 Speed Measurement

To determine the fragments' speed is a difficult task which has two basic options:

Motion capture: high speed cameras can be used to record the event and get the speed by video post processing. This method can be more accurate by means of

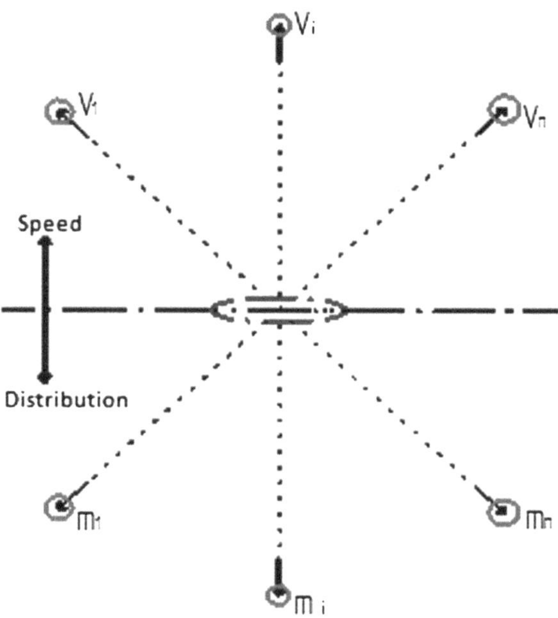

Fig. 2 Symmetry simplification for measurement. Dividing the field by the symmetry axis one can collect only velocity in a half of the arena and mass distribution in the opposite part

Table 1 Arena's radius for different explosive masses

Explossive mass (kg)	Arena radius (m)
To 0.6	2
0.6–10	5
10–130	10

U.S. Army Test and Evaluation Command. ITOP 4-2-813 (1993)

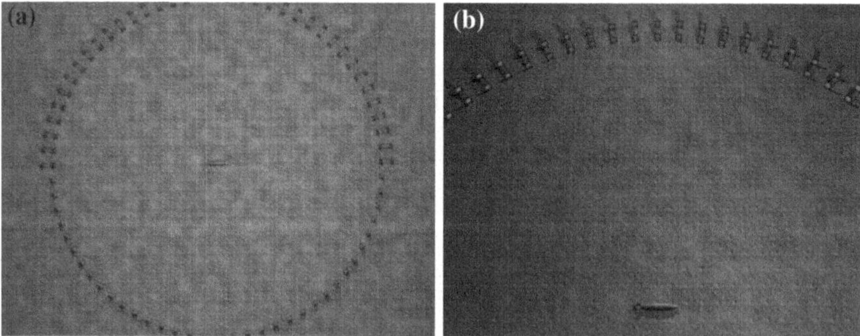

Fig. 3 Render of a test arena (**a**) *Top* view (**b**) Angle view

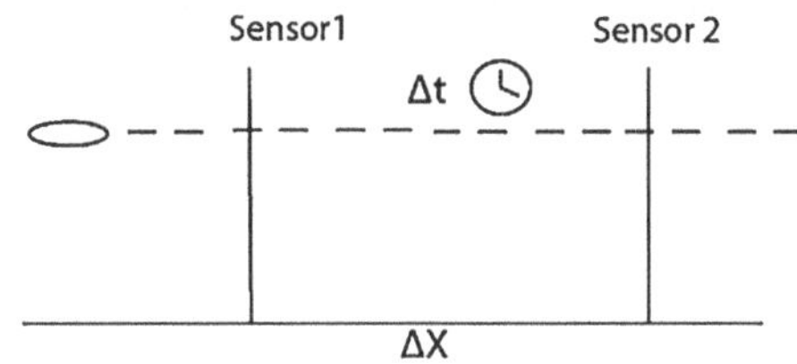

Fig. 4 Sketch of speed measurement by fragment detection

acquiring additional information; on the other hand it could be expensive and lead to complications in the construction of the arena.
Fragment detection: using an electronic system, one can determine the time between two consecutive events of fragment detection, thus obtaining the average speed.

As illustrated in Fig. 4, the principle of fragment's speed determination is timing of two events at a given distance. When sensor1 detects the fragment, a chronometer is activated to count the flight time until the fragment is detected by sensor 2.

Once the time is known, the average speed between sensors can be obtained simply by

$$V_{\text{avg}} = \frac{\Delta X}{\Delta t} \tag{1}$$

Alternatively, instead of sensor 1, the detonation starter can be used to trigger the chronometer and then the average flight time to the arena radius is obtained.

3 Timing Device

Sensor reading and timing was accomplished by a microcontroller driven circuit. It handles the sensors' digital signal through pulled-down inputs and shows the result in a LCD display for the ease of data collecting. The process executed by the circuit is shown in Fig. 5 as a behaviour diagram.

The case where a fragment at a constant speed v is traveling between a pair of sensors is analyzed. When the counter is triggered after destruction of the first sensor, it starts accumulating an integer number (N) at each time step of the digital clock until the destruction of sensor 2. At this moment, the counter stops at a value N. Due to the discrete nature of digital timing, the actual flight time is between N and $N + 1$ clock pulses, thus leading to an uncertainty in the measured speed dependent on the actual speed of the fragment.

$$\frac{f\Delta X}{N} < V_{\text{avg}} < \frac{f\Delta X}{N+1} \tag{2}$$

Knowing that the number of pulses is related to the average speed, the interval of speed uncertainty can be found to be

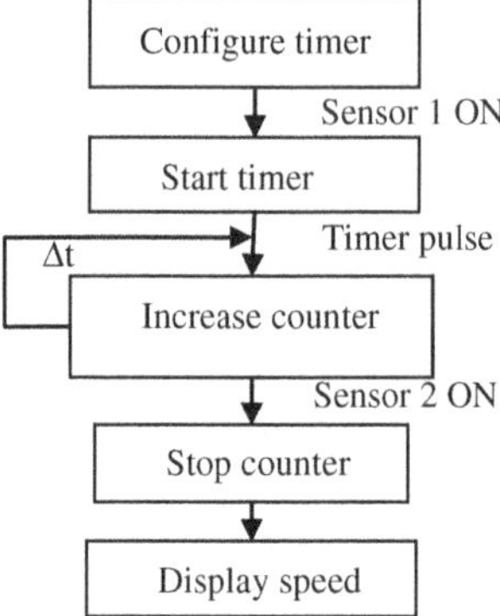

Fig. 5 Circuit behavior diagram

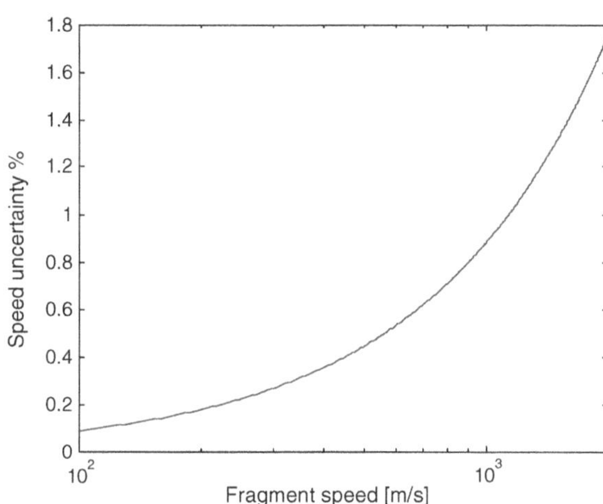

Fig. 6 Speed timing uncertainty dependent on actual speed

$$U_{\mathrm{V}} = \frac{V^2}{f\Delta X + V} \tag{3}$$

Figure 6 shows the uncertainty for the sample case of $x = 1$ m, $f = 112$ kHz.

4 Results

For the sensor, a disposable barrier type of sensor, based on continuity interruption, was developed. It consists of a conductive wire on which a digital high signal is transmitted and gets interrupted by the fragment cutting the wire when it hits the sensor (Fig. 7).

A simple printed circuit board (PCB) manufacturing method was used for producing. Using FR-4 copper clad sheets. The pattern of the sensor was created by photolithography technique (Fig. 8).

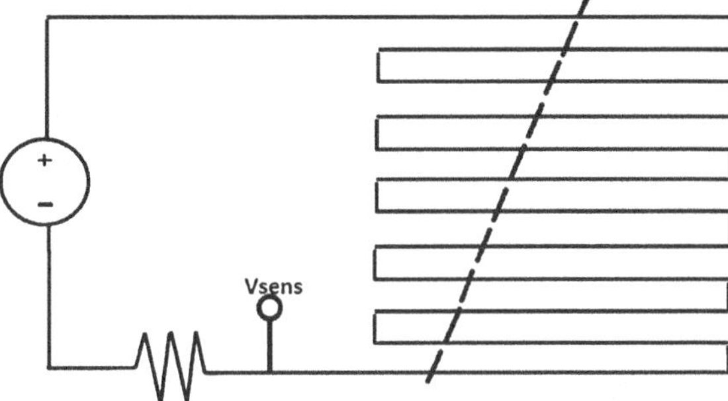

Fig. 7 Sensor operation schematic. When continuity is interrupted low voltage at sense point is detected

Fig. 8 FR-4 Sensor

Laminates are commercially available in virtually any size and thickness varying from 0.1 to 3 mm. This parameter is of important consideration because the quantity of material opposing to the fragment's trajectory can modify its dynamics and affect the measurement. Using the software AUTODYN®, a coupled smoothed particle hydrodynamics (SPH)—Lagrange simulation was implemented to observe the magnitude of speed reduction due to the rupture of sensors of different thickness in fragment's trajectory at the different considered speeds. Simulation condition is presented in Table 2, and a contour map of the velocity magnitude is shown in Fig. 9 for one case of impact speed and thickness of the sensor.

With these conditions, the results show the expected increasing on speed alteration due to the extra thickness of the sensor. These resulting data are presented in Table 3, and plotted for lecture convenience in Fig. 10.

Finally, a small test of the system was executed placing the sensor pair at a distance of 2.5 m from the point of detonation of a 75 g explosive charge. Sensors were effectively affected by small casing fragments of diameter inferior to 2 mm, obtaining the resulting speed of 224 m/s as the average of three executions of the test with registered speeds of 264, 236 and 186 m/s. Dispersion in the results is expected for the complexity of fragmentation process where an statistical analysis

Table 2 Simulation conditions

Fragment	5 mm Dia sphere
Material	Structural steel (A36)
Material properties	Bulk modulus: 200 GPa
	Shear modulus: 90 GPa
	Density: 7,900 kg/m^3
	Yield stress: 200 MPa
	Ultimate strain 0.4
Method	Lagrange
Initial conditions	Initial speed before impact
Sensor	Sheet of defined thickness
Material	Glass reinforced epoxy laminate (FR-4)
Material properties	Young Modulus: 16.5 GPa
	Shear Modulus: 6.85 GPa
	Density: 1,850 kg/m^3
	Ultimate strain 0.1
Method	Smoothed particle hydrodynamics (SPH)
Constraints	No displacement of lateral sides

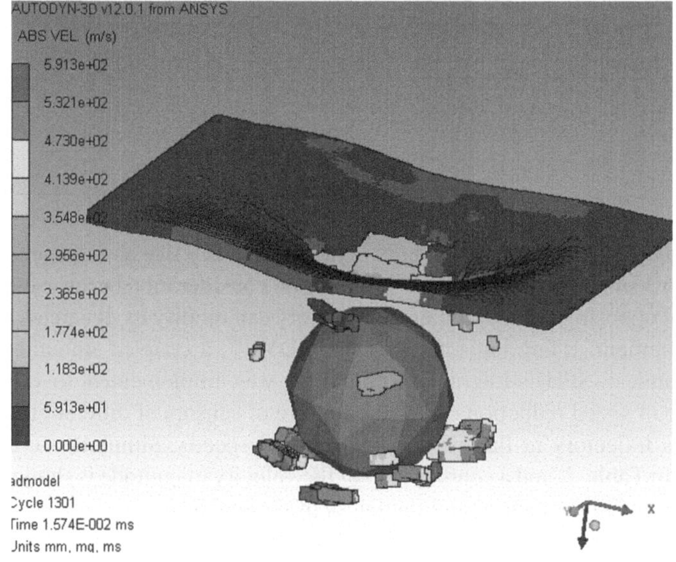

Fig. 9 Contour map of speed for impact of 500 m/s fragment on 0.3 mm thickness sensor

with an increased number of repetitions is required to obtain reliable data for a specific explosive object (Figs. 11 and 12).

The test performed allowed observing the operation of the device on speed measurement, as a preliminary test for an evaluation of a fully instrumented arena as the one depicted in Fig. 3.

Table 3 Results of speed reduction for different simulation conditions

	% Speed reduction	Thickness (mm) 0.15 mm	0.3 mm	0.5 mm
	100	4.1	25.3	46.5
Speed	200	1.6	9.8	15.6
(m/s)	300	1.0	3.9	6.5
	400	0.8	2.9	4.9
	500	0.7	2.2	4.1
	1000	0.7	1.6	2.9
	1500	0.7	1.5	2.6

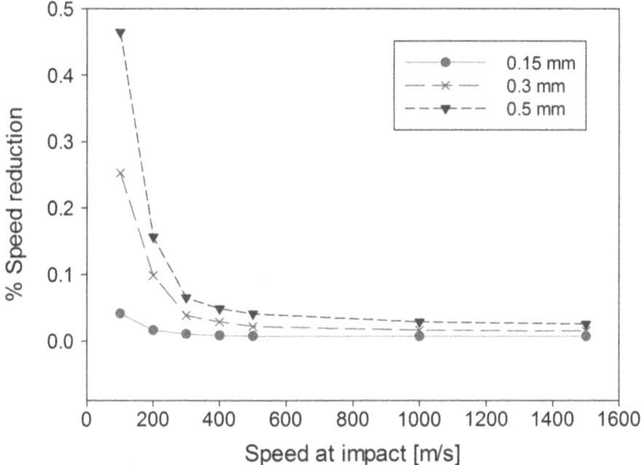

Fig. 10 Speed reduction for different simulation conditions

Fig. 11 System test

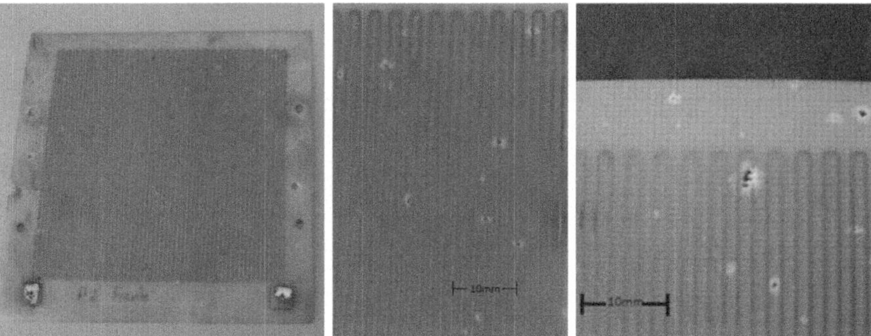

Fig. 12 Sensor after test

5 Conclusions

A methodology for testing the effect of explosion, focused on fragmentary weapons, was developed to acquire important information of fragment distribution and damage capability. In this method, a continuity interruption barrier sensor was designed to be made using a PCB fabrication method. This type of sensor is simple to manufacture and can be modified to detect fragments of a particular size by changing the width and spacing of the cooper tracks.

Microcontroller driven circuit accomplished the timing strategy for sensor events; this approach allows reproduction of the system, thus it makes a possible choice for executing experimentation in large scale due to the simplicity and low cost of the instrumentation.

Although other types of sensors (e.g. optical) will not alter the fragment's speed, their use is restricted only to contained experiments where trajectory is controlled. In other cases, the use of disposable sensors, like the one presented, is preferred. The use of elements like high speed cameras increases the test cost, including difficulties of installation and equipment protection.

Through computer simulation, speed reduction at different conditions can be estimated to give an approximation to the level of intrusion of the method in the measurement; from the results it was evident that this is the most influent cause of uncertainty, decreasing the speed in the order of 5 % for sensor thickness under 0.3 mm affected by fragments in a range of speed over 500 m/s.

References

1. Zigler, A., Ludmirsky, A., et al.: Laser-generated shock-wave velocity measurements using a visible backlighting technique. J. Phys. E Sci. Instrum. **19**, 309–315 (1986)
2. Anderson, B.W., Settles, G.S., et al.: High-speed imaging of shock-wave motion in aviation security research. American Physical Society, 54th Annual Meeting of the Division of Fluid Dynamics (2001)

3. Wu, J., Liu, J., et al.: Effect of shock wave on fabricated anti-blast wall and distribution law around the wall under near surface explosion. Trans. Tianjin Univ. **14**, 514–518 (2008)
4. Eisler, R.D., Chalterjee, A.K., et al.: Simulation and modeling of penetrating wounds from small arms. Stud. Health Technol. Inform. **29**, 511–522 (1996)
5. DuBay, D., Bir, C.A., et al.: Biomechanics and injury risk assessment of less lethal munitions: analysis of the defense technology #23BR bean bag. Defense Technology Corporation of America and Institute for Preventative Sports Medicine. St. Joseph Mercy Hospital, Ann Arbor (1995)
6. Jussila, J.: Wound ballistic simulation: assessment of the legitimacy of law enforcement firearms ammunition by means of wound ballistic simulation Helsinki. http://ssf1910.dk/document/info/balistik.pdf (2005). Accessed 20 Oct 2010
7. US Army Test and Evaluation Command :Static testing of high explosive munitions for obtaining fragment space distribution. ITOP 4-2-813 (1993)

Forward Modeling of Seabed Logging by Finite Integration and Finite Element Methods

Noorhana Yahya, Majid Niaz Akhtar, Nadeem Nasir, Hanita Daud and Marneni Narahari

Abstract Seabed electromagnetic (EM) modeling for detection of deep target hydrocarbon reservoirs has been a challenge for oil and gas industry. More precise and accurate electromagnetic (EM) methods are required for better detection of hydrocarbon (HC) reservoirs. To overcome this problem, Finite integration method (FIM) and Finite element method (FEM) were chosen for 3D modeling of seabed logging to produce more precise EM response from the hydrocarbon reservoir. EM modelling is used to investigate the total electric and magnetic fields instead of scattered electric and magnetic fields, because it shows accurate and precise resistivity contrast at the target depth of up to 3000 m below seafloor. The FIM and the FEM were applied to our proposed seabed model having an area of 20×20 km. It was observed that the FIM showed 6.52 % resistivity contrast at a target depth of 1000 m whereas the FEM showed 16.78 % resistivity contrast at the same target depth for the normalised E-field. It was also found that normalised

N. Yahya (✉) · H. Daud · M. Narahari
Department of Fundamental and Applied Sciences,
Universiti Teknologi PETRONAS, Bandar Seri Iskandar,
31750 Tronoh, Perak Malaysia
e-mail: noorhana_yahya@petronas.com.my

H. Daud
e-mail: hanita_daud@petronas.com.my

M. Narahari
e-mail: marneni@petronas.com.my

M. N. Akhtar · N. Nasir
Department of Electrical and Electronic Engineering,
Universiti Teknologi PETRONAS, Bandar Seri Iskandar,
31750 Tronoh, Perak Malaysia
e-mail: majidniazakhtar@gmail.com

N. Nasir
e-mail: nadeemntu@hotmail.com

A. Öchsner et al. (eds.), *Design and Analysis of Materials and Engineering Structures*, Advanced Structured Materials 32, DOI: 10.1007/978-3-642-32295-2_11,

E-field response decreased as the target depth increased gradually by 500 m from 1000 to 3000 m at constant frequency of 0.125 Hz and current of 1250 A. It was also observed that at frequency of 0.125 Hz, phase versus offset (PVO) showed 3.8 % for FIM whereas 6.58 % for FEM better delineation of hydrocarbon at 3000 m target depth. PVO of electric field gives better delineation of HC presence compared to magnitude of E and H fields.

1 Introduction

Currently, oil and gas industry is facing challenge for detecting deep target hydrocarbon reservoirs at 3000 m from seabed [1–4]. In seabed logging (SBL), a horizontal electric dipole (HED) antenna is used to diffuse low frequency electromagnetic signals in the marine environment for deep target exploration. Dipole antenna is towed at 30 m above the sea floor for the detection of hydrocarbon [5–8]. An electromagnetic (EM) wave after reflected, refracted and guided from the hydrocarbon reservoir is detected by electric or magnetic field sensors placed on the seabed [9–12]. The EM waves captured by the receivers are to be processed to investigate the presence of hydrocarbon. In this chapter, we will use the Computer Simulation Technology (CST) software for the FIM. The CST is used to discretize each Maxwell's equations at low frequency to investigate the resistivity contrast. For FEM, Comsol Multiphysics software is used as a tool for low frequency problems. Maxwell's equations used in the CST software are given below.

$$\oint_{\partial A} \vec{E}.\vec{ds} = -\frac{\partial}{\partial t}\iint_A \vec{B}.\vec{dA} \tag{1}$$

$$\oint_{\partial A} \vec{H}.\vec{ds} = \iint_A [\frac{\partial \vec{D}}{\partial t} + \vec{j}].\vec{dA} \tag{2}$$

$$\oiint_{\partial V} \vec{B}.\vec{dA} = 0 \tag{3}$$

$$\oiint_{\partial V} \vec{D}.\vec{dA} = Q \tag{4}$$

FEM has become a powerful tool to solve Controlled Source Electromagnetic (CSEM) problems and a versatile method for finding approximate solutions of partial differential equations (PDE) as well as integral equations. In FEM, Maxwell's equations are also used to solve electromagnetic (EM) problems. It was found that FEM generates detailed visualization of resistive layers in marine control source electromagnetic (MCSEM) environment [13]. FEM uses Galerkin method which minimizes the errors over the entire volume of the proposed model. The process of discretization of the region by creating meshes was done within the finite element approach [14]. FEM discretizes the region by creating meshes on the proposed model. These meshes subdivide a large geometry into a number of non overlapping sub regions. It is important to discretize regular or irregular structures into a finite number of degrees of freedom (DOF). Due to unstructured meshes, the FEM can be easily applied to the arbitrarily geological structure. Finite element (FE) modeling gives results with less than 1 % error as compared to the Finite Difference Time Domain (FDTD), finite integration (FI) and Method of Moment (MoM) methods [15].

Finite element modeling of geophysical electromagnetic structures has been widely studied [16, 17]. The CSEM forward seabed one-dimensional model uses primary conductivity and secondary fields (E_S and H_S) due to inhomogeneities in the CSEM environment. The equations for electric and magnetic fields are given below [18]:

$$\nabla \times E^s = i\omega\mu_0 H^s \tag{5}$$

$$\nabla \times H^s - \sigma E^s = \sigma_s E^p \tag{6}$$

Equations (5), (6) represent the 3D electric and magnetic field equations due to the resistivity (conductivity) changes in 2D environment. The electric and magnetic field equations to find the variables E_x, E_y, E_z and H_x, H_y, H_z components in the marine environment are

$$\frac{\partial E_z^s}{\partial y} - \frac{\partial E_y^s}{\partial z} = i\omega\mu_0 H_x^s \tag{7}$$

$$\frac{\partial E_x^s}{\partial y} - ik_x E_x^s = i\omega\mu_0 H_y^s \tag{8}$$

$$ik_x E_y^s - \frac{\partial E_x^s}{\partial y} = i\omega\mu_0 H_z^s \tag{9}$$

$$\frac{\partial H_z^s}{\partial y} - \frac{\partial H_y^s}{\partial z} - \sigma E_x^s = \sigma_s E_x^p \tag{10}$$

$$\frac{\partial H_z^s}{\partial y} - ik_x H_z^s - \sigma E_y^s = \sigma_s E_y^p \tag{11}$$

$$ik_x H_y^s - \frac{\partial H_x^s}{\partial y} - \sigma E_z^s = \sigma_s E_z^p \tag{12}$$

Where, k_x is the wave number. Equations (7)–(12) are used to find the 3D solution of the electric and magnetic field response from the seabed environment.

This chapter describes the presence of HC by using FIM and FEM techniques. We used constant geological strata layers and fixed the source to be operating at frequency of 0.125 Hz with current of 1250 A (Fig. 1a). The magnitude and phase of the electric and magnetic fields were evaluated with and without the presence of HC.

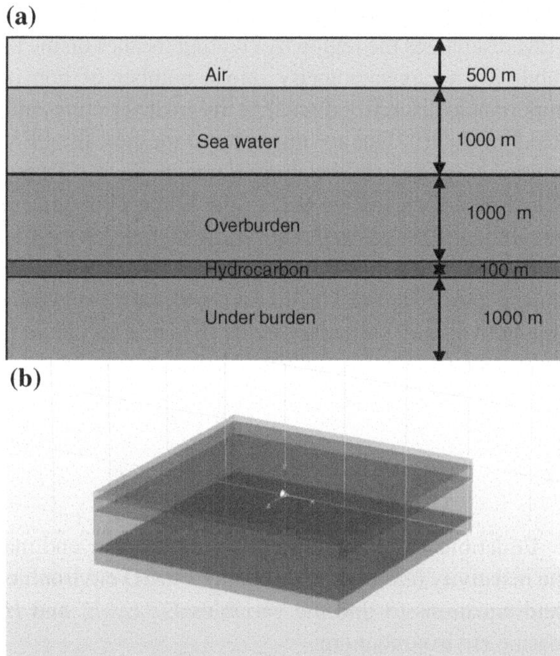

Fig. 1 (**a**) Schematic diagram of the proposed model and (**b**) CST simulated model

2 Methodology

2.1 Modelling of Seabed Logging by Computer Simulated Software

CST software was used for FIM to detect deep target hydrocarbon between 1000–3000 m underneath the seabed. The model area was assigned as 20 × 20 km and replicate the real seabed environment with various target positions. Environment with and without hydrocarbon were also prepared for comparison purposes later. There were few steps involved in generating the CST simulated models. The first step was to set parameters for aluminium antenna. In this case, antenna length was set as 270 m, frequency of 0.125 Hz and current of 1250 A. The second step was to set the parameters for the model. The air thickness was set as 500 m, sea water depth as 1000 m, overburden thickness as 1000 m, hydrocarbon thickness as 1000 m and under burden thickness as 1000 m. Overburden thickness was increased as the target depth was varied gradually (every 500 m) from 1000 to 3000 m. The third step was to apply electric boundary conditions (Table 1). The fourth step was to run the low frequency full wave solver to simulate the sea bed model. The final step was post processing to generate the simulated data for results analysis at different target depths. Maxwell's equations for magnetic and electric

Table 1 Model parameters of target depth, air, sea water and under burden

Target depth (m)	Air thickness (m)	Under burden (m)	Hydrocarbon thickness (m)	Sea water depth	Frequency (Hz)
1000	500	1000	100	1000	0.125
1500	500	1000	100	1000	0.125
2000	500	1000	100	1000	0.125
2500	500	1000	100	1000	0.125
3000	500	1000	100	1000	0.125

Table 2 Relative permittivity (ε_r) and conductivity (σ) values of air, sea water overburden/under burden and hydrocarbon

	Air	Sea water	Under burden/ overburden	Hydrocarbon
ε_r	1.006	80	30	4
σ (S/m)	$1.0e^{-3}$	3	1.5	0.001

fields were used in the software codings to get the electric and magnetic field response with and without hydrocarbon (HC). A schematic diagram of the proposed seabed model with the CST simulated model is as shown in Fig. 1.

2.2 *Modeling of Seabed Logging by Finite Element Method*

The Comsol Multiphysics software was used for the FEM with the same parameters set up as in FIM using the CST software. Triangular meshes with a free mesh parameter of 1.0e-3 were used to simulate the seabed model. More refined meshes were also created to get more accurate and precise results. Sub domain settings with relative permittivity, conductivity of air, sea water, under burden, overburden, hydrocarbon and conductivity were adjusted, as given in Table 1. Partial differential equations for magnetic and electric fields in the wave number domain were selected to get the total electric and magnetic field response with and without presence of hydrocarbon. Table 2 shows the relative permittivity and conductivity values of air, sea water overburden, under burden and hydrocarbon. The seabed logging model with meshes generated by FEM is as shown in Fig. 2.

3 Results and Discussion

3.1 *3D Seabed Modeling Results from CST (Computer Simulated Software)*

The simulated data after the post processing proposed seabed model was normalised. Data was normalised and locked to see the response of the EM signal with

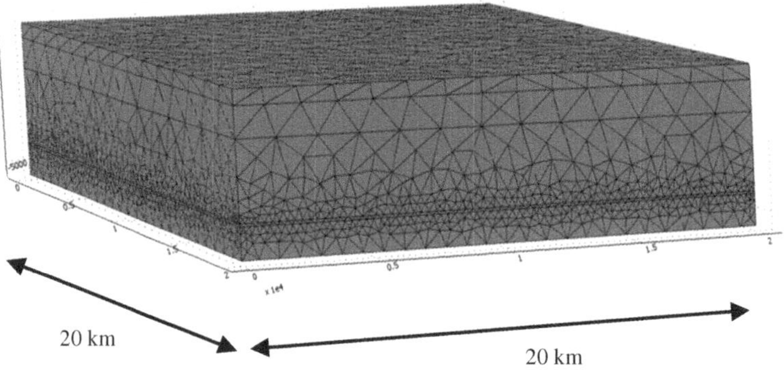

Fig. 2 Seabed logging model with meshes generated by FEM

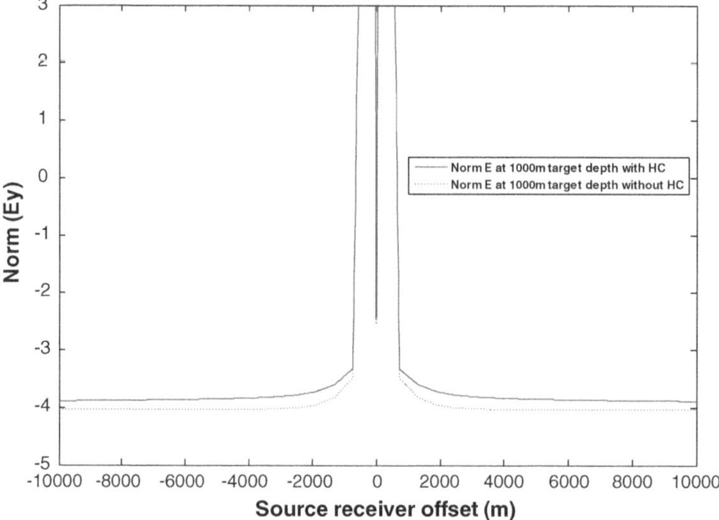

Fig. 3 Normalised E-field versus source receiver offset with and without HC at 1000 m target depth using CST/FIM

and without the presence of hydrocarbon. The formula used to normalise that data was,

$$z = \frac{X - X_{ave}}{X_{Std}} \tag{13}$$

Whereas, X is the simulated data, X_{ave} represents the average data and X_{Std} is the standard deviation of data.

Figures 3, 4, 5, 6, and 7 show the normalised E-field response with and without hydrocarbon located at 1000, 1500, 2000, 2500 and 3000 m respectively

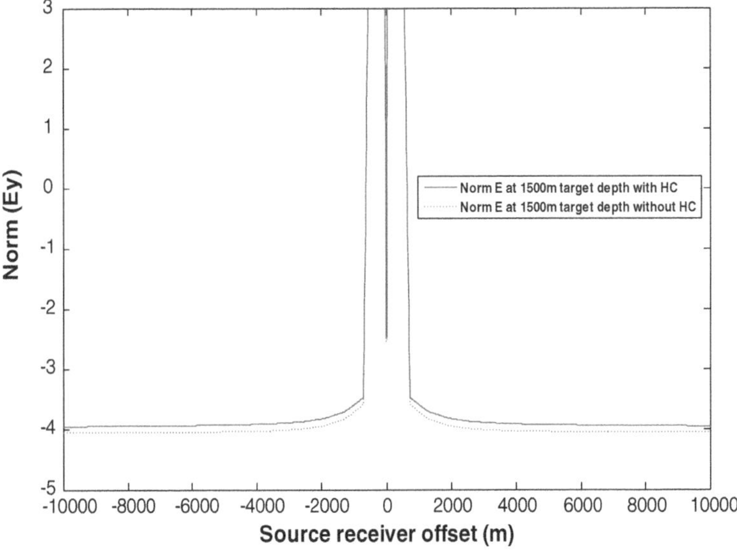

Fig. 4 Normalised E-field versus source receiver offset with and without HC at 1500 m target depth using CST/FIM

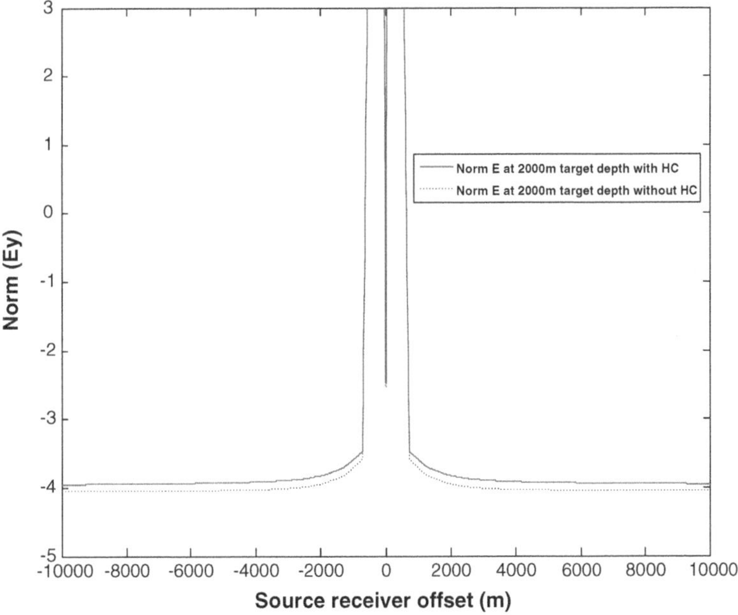

Fig. 5 Normalised E-field versus source receiver offset with and without HC at 2000 m target depth using FIM/CST

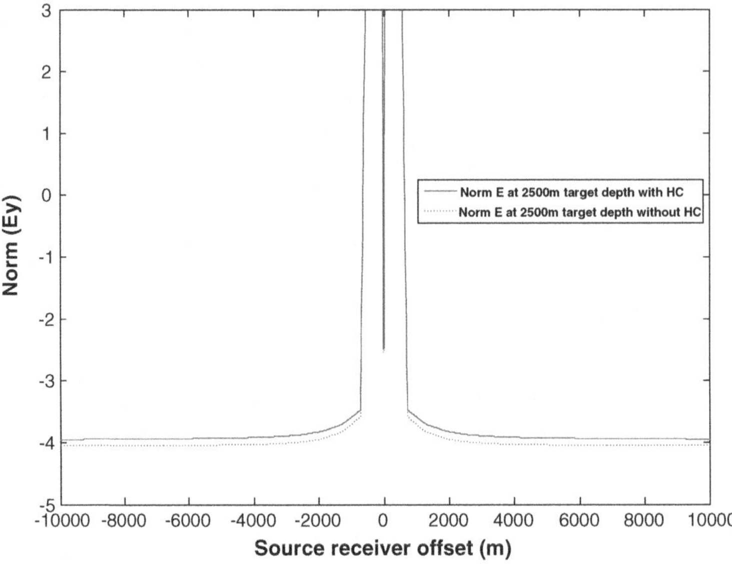

Fig. 6 Normalised E-field versus source receiver offset with and without HC at 2500 m target depth using CST/FIM

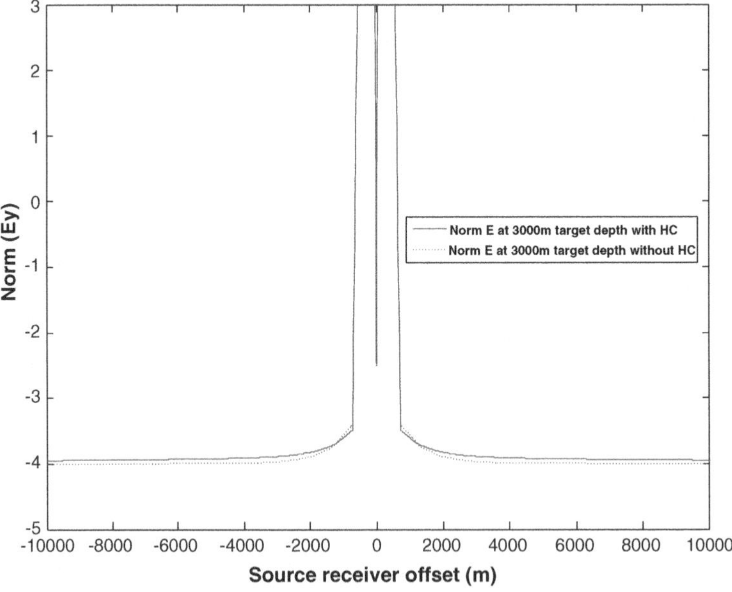

Fig. 7 Normalised E-field versus source receiver offset with and without HC at 3000 m target depth using CST/FIM

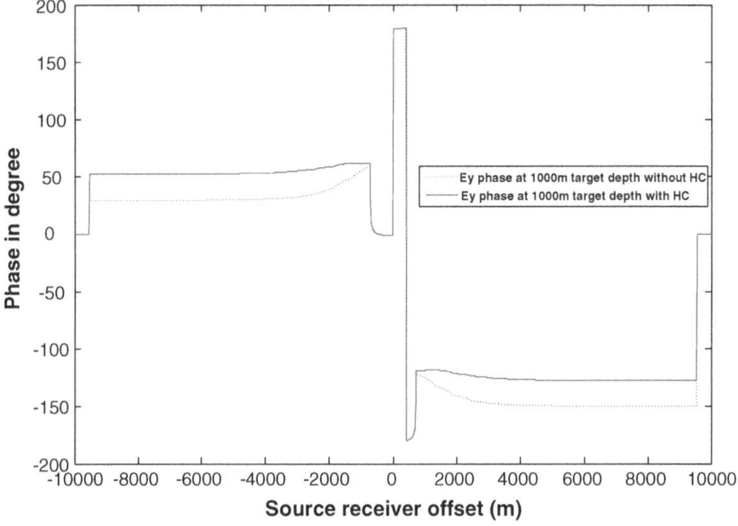

Fig. 8 PVO with and without HC at 1000 m target depth using FIM

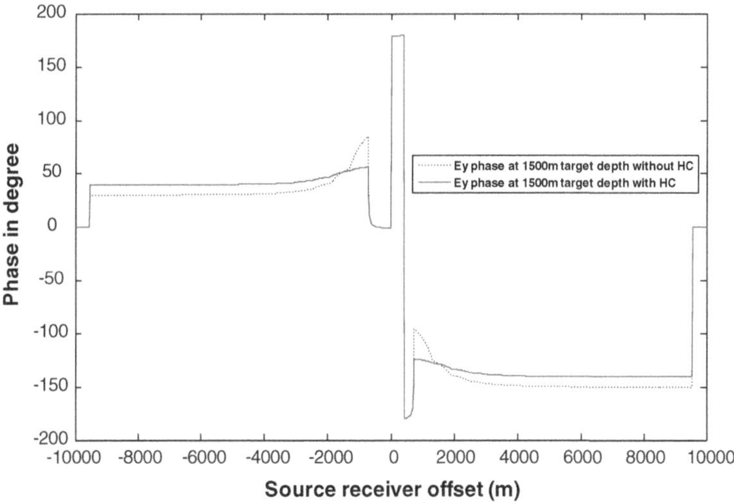

Fig. 9 PVO with and without HC at 1500 m target depth using FIM

underneath the seafloor. At 1000 m target depth the normalised E-field with and without hydrocarbon shows clear differences at the offset of the 20 km, as shown in Fig. 3. Normalization of the data was done to see the difference at the offset between with and without hydrocarbon at far offset [19]. It is observed that the E-

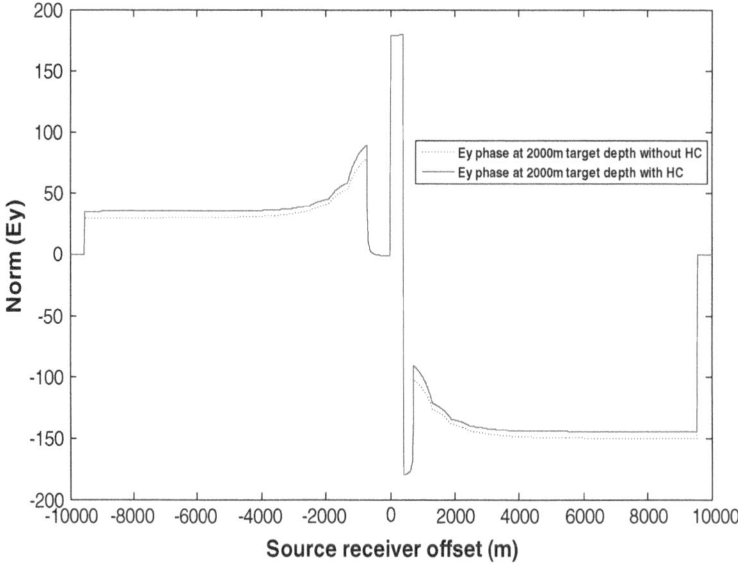

Fig. 10 PVO with and without HC at 2000 m target depth using FIM

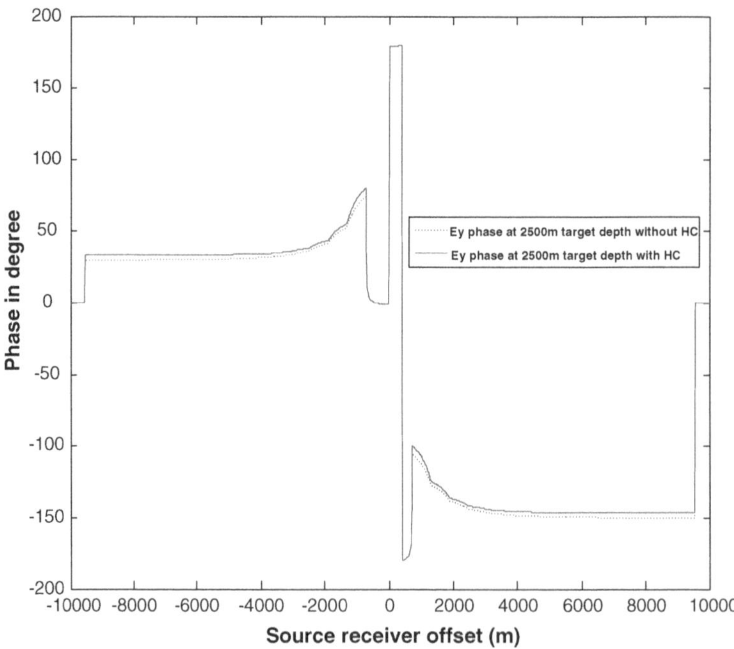

Fig. 11 PVO with and without HC at 2500 m target depth using FIM

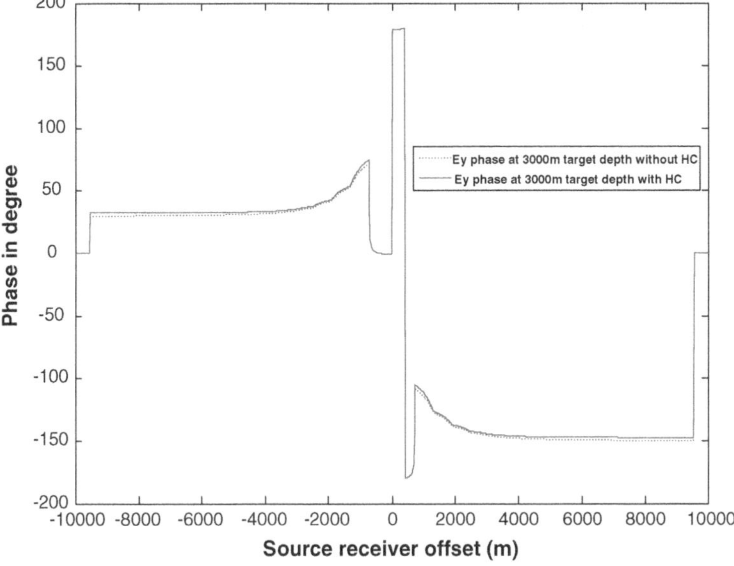

Fig. 12 PVO with and without HC at 3000 m target depth using FIM

Table 3 Percentage difference between MVO and PVO with and without HC at different target depths by using CST software

Target depth (m)	Transmitter current (A)	Frequency (Hz)	Average % difference in normalised Ey field with and without HC	Average % difference in Ey phase with and without HC
1000	1250	0.125	6.52	62.66
1500	1250	0.125	2.47	14.78
2000	1250	0.125	2.18	10.86
2500	1250	0.125	1.8	6.09
3000	1250	0.125	1.61	3.8

field with and without hydrocarbon (at 1000 m target depth) gave a 6.52 % difference. The electric fields response are 2.47, 2.18, 1.8, and 0.61 % respectively if compared to without hydrocarbon layer at target depth of 1500, 2000, 2500, and 3000 m, respectively as shown in Figs. 4, 5, 6, and 7. The fall of E-field response is due to the weakness of E-field to give better HC delineation.

Phase versus offset (PVO) data is important to show the delineation of HC. The presence of HC is indicated as phase leads where as phase lag indicate no HC. At the edge of the source receiver offset or the receiver position outside the reservoir boundary, the phase response dies out rapidly with the distance from the edge [20–22]. At 1000 m target depth, average phase advances 62.66 % of the value than without hydrocarbon. Similarly, the phase lags 14.78, 10.86, 6.09 and 3.8 %

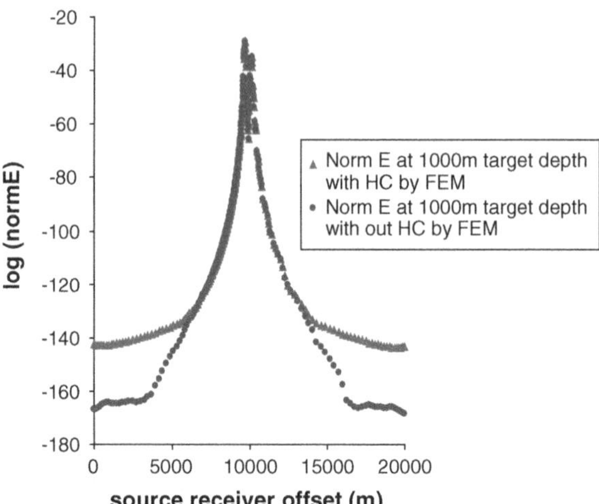

Fig. 13 Magnitude of E-field versus source receiver offset with and without HC at 1000 m target depth using FEM

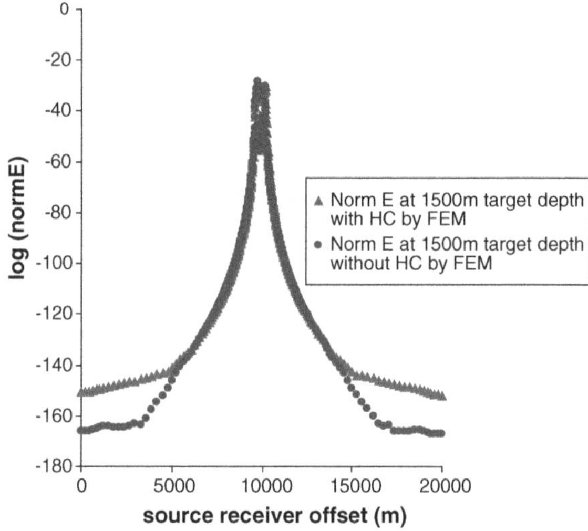

Fig. 14 Magnitude of E-field versus source receiver offset with and without HC at 1500 m target depth using FEM

respectively with hydrocarbon than those without hydrocarbon, as shown in Figs. 8, 9, 10, 11, and 12. Table 3 shows the average percentage difference of normalised Ey field and phase with and without HC at various target depth.

3.2 3D Finite Element Modeling Results

FEM using Comsol Multiphysics has been applied to simulate our proposed model. Finite element (FE) creates meshes on the proposed model to get accurate and precise EM response from the marine environment such as overburden, under

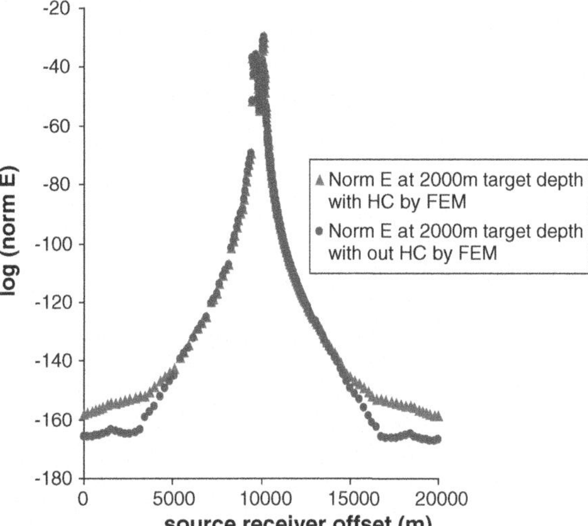

Fig. 15 Magnitude of E-field versus source receiver offset with and without HC at 2000 m target depth using FEM

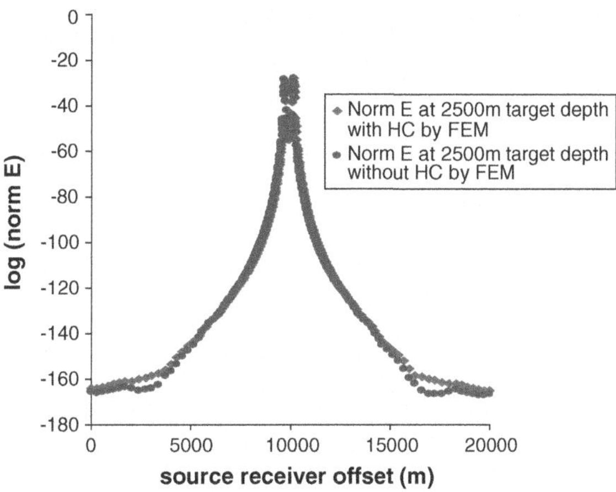

Fig. 16 Magnitude of E-field versus source receiver offset with and without HC at 2500 m target depth using FEM

burden, seawater, and hydrocarbon etc. It was found that finite element software has an ability to create more refined meshes with thousands of interconnecting nodes on any proposed seabed model. It was observed that more refined meshes at the transmitter and receiver especially small if transmitter and receiver offsets were used to get more precise and accurate results [23]. We used the normalised E-field to get better delineation of hydrocarbon (HC) beneath the seabed by using the equation given below

$$20 \times \log_{10}(NormalisedE) \tag{14}$$

Figures 13, 14, 15, 16, and 17 show the normalised E-field response with and without hydrocarbon (HC) located at 1000, 1500, 2000, 2500 and 3000 m below

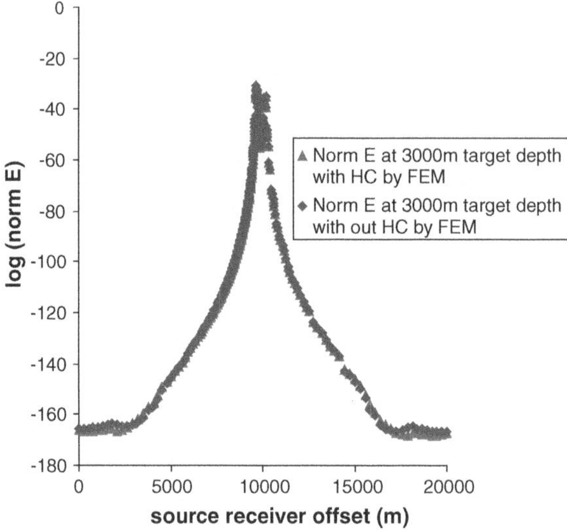

Fig. 17 Magnitude of E-field versus source receiver offset with and without HC at 3000 m target depth using FEM

Table 4 Comparison of normalised E-field with and without hydrocarbon with different target depths

Target depth (m)	Frequency (Hz)	Current (A)	% difference in normalised E-field with and without hydrocarbon
1000	0.125	1250	16.78
1500	0.125	1250	10.59
2000	0.125	1250	5.690
2500	0.125	1250	1.821
3000	0.125	1250	0.601

the seafloor respectively. It is found that the normalised E-field at the offset decreases as the depth increases from 1000 to 3000 m (Figs. 13, 14, 15, 16, and 17). It can be seen that at target depth of 1000 m normalised E-field response was 16.78 % but as the depth increased up to 3000 m, the E-field response decreased up to 0.601 %. Comparison of percentage of normalised E-field with and without hydrocarbon with different target depths are given in Table 4.

Phase response (PVO) with the source receiver offset is shown in Figs. 18, 19, 20, 21, and 22. It was observed that the percentage difference in phase also decreases as the target depth increases from 1000 to 3000 m, as shown in Figs. 18, 19, 20, 21, and 22. It is found that the phase at 1000 m gives 69.5 % difference with and without

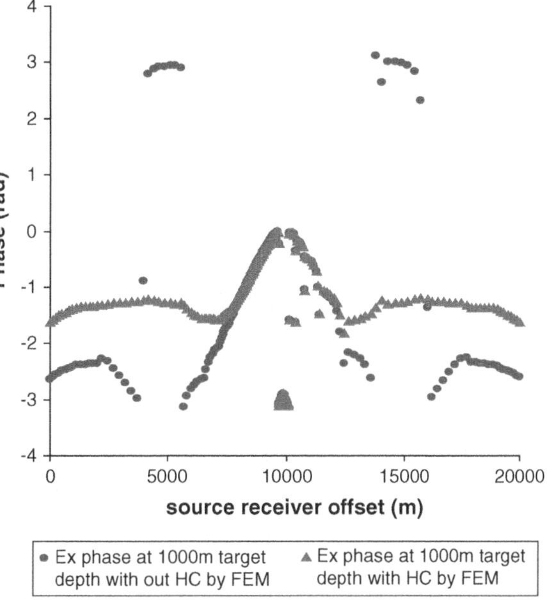

Fig. 18 PVO with and without HC at 1000 m target depth using FEM

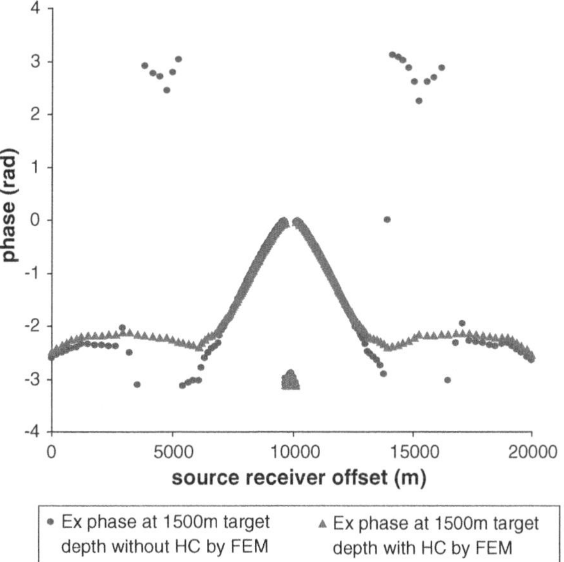

Fig. 19 PVO with and without HC at 1500 m target depth using FEM

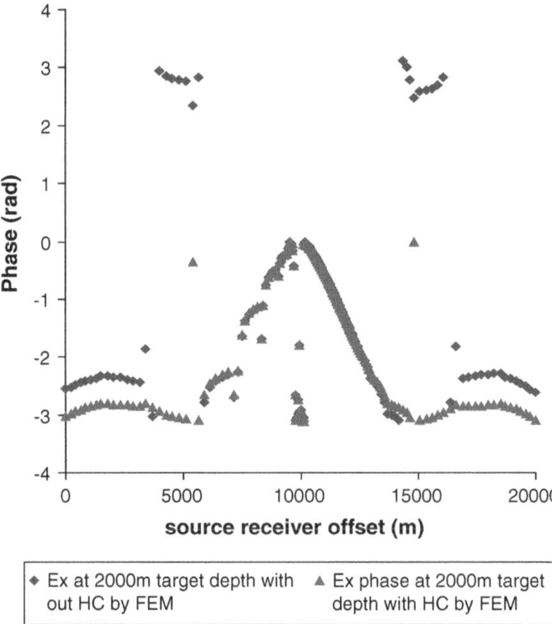

Fig. 20 PVO with and without HC at 2000 m target depth using FEM

hydrocarbon (Fig. 18) whereas at 3000 m target depth delineation is 6.58 % (Fig. 22) [22]. Table 5 shows the comparison between phase with source receiver offset with and without hydrocarbon at target depth of 1000, 1500, 2000, 2500 and 3000 m respectively. It is observed that as the target depth increases from 500 to 3000 m, the electromagnetic signal attenuates rapidly. Figure 17 shows that at a 3000 m target depth, the EM signal received by the receivers detects the same response for with and without the presence of hydrocarbon.

Comparative study of magnitude versus offset (MVO) and phase versus offset (PVO) is also done (Figs. 21, 22). It was found that with the presence of hydrocarbon the phase versus offset (PVO) shows continuous response whereas without hydrocarbon, the response is discontinuous. It is observed that the magnitude versus offset (MVO) shows a difference of 1.82 % at a target depth of 2500 m. On the other hand, the phase versus offset (PVO) response shows a difference of 18.1 % with and without the presence of hydrocarbon. Similarly, for 3000 m target depth, we cannot see the difference between with and without hydrocarbon for magnitude versus offset (MVO) response whereas from phase versus offset (PVO), we are able to see the difference between with and without presence of hydrocarbon (Fig. 22). This indicates the importance of PVO as a better delineation exercise for both FIM and FEM techniques.

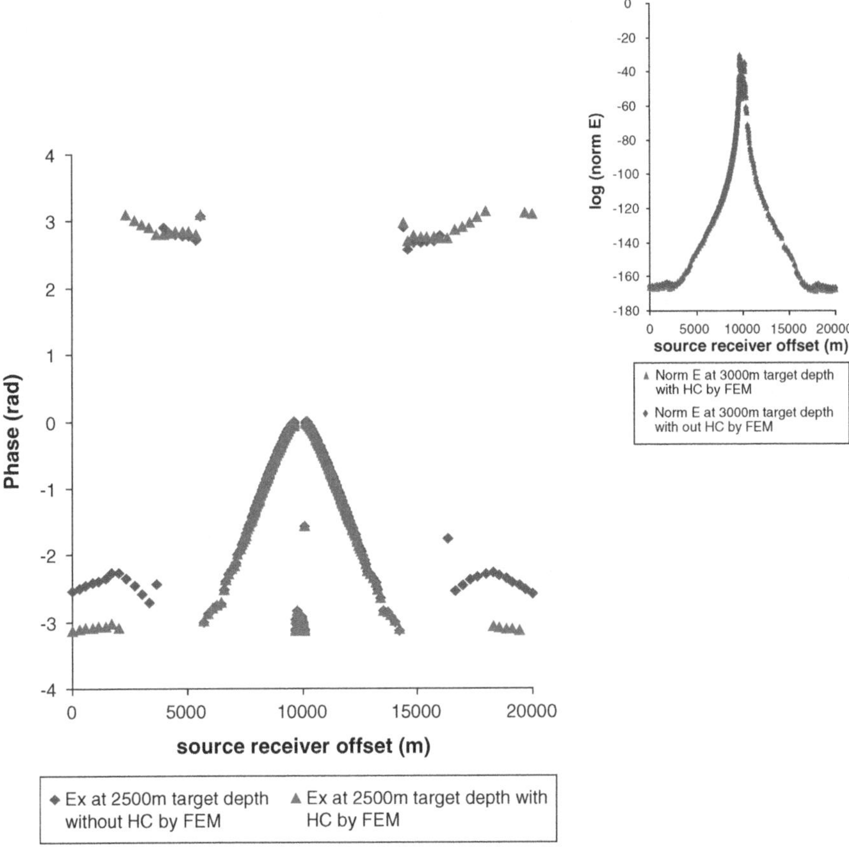

Fig. 21 PVO with and without HC at 2500 m target depth using FEM. (Note: Fig. 16 is superimposed to see the difference between MVO and PVO for HC delineation at 2500 m target depth)

4 Conclusion

FIM and FEM have been successfully applied to get more precise EM response from hydrocarbon reservoirs. It was shown that FIM shows a 6.52 % resistivity contrast at a target depth of 1000 m whereas FEM shows a 16.78 % resistivity contrast of the normalised E-field. It was also observed that the normalised E-field response decreases as the target depth increases from 1000 to 3000 m with a phase difference of 3.8 % for FIM and 6.58 % for FEM techniques. We also concluded that the PVO has better HC delineation compared to the MVO for E-field and frequency of 0.125 Hz and current of 1250 A are not suitable for 3000 m target depth below seabed.

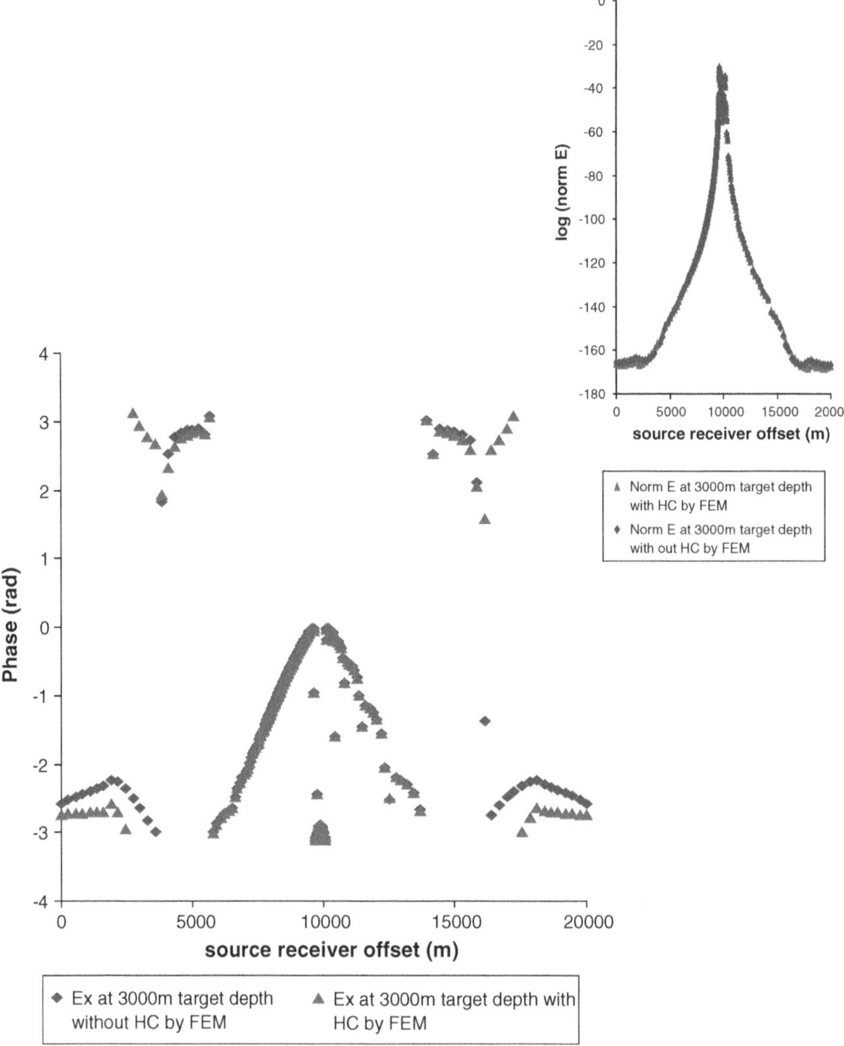

Fig. 22 PVO with and without HC at 3000 m target depth using FEM (Note: Fig. 17 is superimposed to see the difference between MVO and PVO for HC delineation at 3000 m target depth)

Table 5 Comparison of PVO with and without hydrocarbon at different target depths

Target depth (m)	Frequency (Hz)	Current (A)	% difference in phase with source receiver offset with and without hydrocarbon
1000	0.125	1250	69.5
1500	0.125	1250	40.2
2000	0.125	1250	22.3
2500	0.125	1250	18.1
3000	0.125	1250	6.58

References

1. MacGregor, L., Sinha, M.: Use of marine controlled-source electromagnetic sounding for sub-basalt exploration. Geophys. Prospect. **48**, 1091–1106 (2000)
2. Tossman, B., Thayer, D., Swartz, W.: An underwater towed electromagnetic source for geophysical exploration. IEEE J. Oceanic Eng. **4**, 84–89 (1979)
3. Yahya, N., Akhtar, M.N., Nasir, N., Shafie, A., Jabeli, M.S., Koziol, K.: CNT Fi–bres/ Aluminium–NiZnFe2O4 based EM transmitter for improved magnitude versus offset (MVO) in a scaled marine environment. J. Nanosci. Nanotechnol. (2011, in Press)
4. Cox, C.S., Constable, S.C., Chave, A.D., Webb, S.C.: Controlled source electromagnetic sounding of the oceanic lithosphere. Nature **320**, 52–54 (1986)
5. Ellingsrud, S., Eidesmo, T., Sinha, M.C., MacGregor, L.M., Constable, S.C.: Remote sensing of hydrocarbon layers by seabed logging (SBL): results from a cruise offshore Angola. Leading Edge **20**, 972–982 (2002)
6. Sinha, M.C., Patel, P.D., Unsworth, M.J., Owen, T.R.E., MacCormack, M.G.R.: An active source electromagnetic sounding system for marine use. Marine Geophys Res **12**, 29–68 (1990)
7. Løseth, L.O., Pedersen, H.M., Pettersen, S., Ellingsrud, T.S., Eidesmo, T.: A scaled experiment for the verification of the seabed logging method. J. Appl. Geophys. **64**, 47–55 (2008)
8. Akhtar, M.N., Yahya, N., Daud, H., Shafie, A., Zaid, H.M., Kashif, M., Nasir, N.: Development of EM wave guide amplifier potentially used for seabed logging. J. App. Sci **1** (2011)
9. Webb, S.C., Constable, S.C., Cox, C.S., Deaton, T.K.: A seafloor electric field instrument. J. Geomagnet. Geoelectric. **37**, 1115–1129 (1985)
10. Chave, A.D., Constable, S.C., Edwards, R.N.: Electrical exploration methods for the seafloor. In: Electromagnetic Methods in Applied Geophysics, SEG, pp. 931–966 (1982)
11. Unsworth, M.J.: Exploration of mid-ocean ridges with a frequency domain electro-magnetic system. Geophys. J. Int. **116**, 447–467 (1994)
12. Webb, S.C., And Cox, C.S.: Electromagnetic fields induced at the seafloor by Raleigh-Stoneley waves. J. Geophys. Res. **87**, 4093–4102 (1982)
13. Sugeng, T., Raiche, A., Wilson, R.: An efficient compact finite element modelling method for practical 3D inversion of electromagnetic data from high contrast complex structures. IAGA WG 1,2 on Electromagnetic induction in Earth, Extended Abstract 18th workshop EL Vendrell, Spain, 17–23 Sept 2006
14. Sadiku, M.N.O.: Numerical methods in Electromagnetics, 2nd edn. CRC Press, Boca Raton (2001)
15. Park, J., Bjornara, T.I., Westerdahl, H., Gonzalez, E.: On boundary conditions for CSEM finite element modeling. In: Proceedings of the Comsol Conference Hannover (2008)
16. Kong, F.N., Johnstad, S.E., Røsten, T., Westerdahl, H.: A 2.5D finite-element-modeling difference method for marine CSEM modeling in stratified anisotropic media. Geophysics **73**(1), F9–F19 (2008)
17. Colin, G.F.: Numerical modeling for geophysical electromagnetic methods, memorial university of Newfoundland, St. Johns.Nl, Canada (2009)
18. Li, Y., Key, K.: 2D marine controlled-source electromagnetic modeling: Part 1 an adaptive finite-element algorithm. Geophysics **72**(2), WA51–WA62
19. Rune, M.: Normalized amplitude ratios for frequency domain CSEM in very shallow water. First Break **26**, 47–53 (2008)
20. Mittet, R., Aakervik, O., Jensen, H., Ellingsrud, S., Stovas, A.: On the orientation and absolute phase of marine CSEM receivers. Geophysics **72**, F145 (2007)
21. Gelius, L.J.: Multi-component processing of sea bed Logging data. PIERS ONLINE **2**(6), 589–593 (2006)
22. Eidesmo, T.: Sea bed logging (SBL), a new method for remote and direct identification of hydrocarbon filled layers in deepwater areas. First Break **20**, 144–152 (2002)
23. King, J.D.: Using a 3D finite element forward modelling code to analyze resistive structures with controlled source electromagnetics in a marine environment. Master's thesis, Dec 2004

3D Mesh Extraction for Transmission Line Matrix Modelling

Alexandre S. Brandão, Fabiana R. Leta and Edson Cataldo

Abstract The ModaVox program was developed to study acoustic propagation of vowel sounds in a 3D vocal tract by using the transmission line matrix (TLM) method. The name ModaVox stands for "Modelador da Voz" in Portuguese, or Voice Modeler in English. At the present moment, it is able to build and solve 3D TLM numerical models of the vocal tract. The meshes are constructed over the voxels (volumetric picture elements) in segmented medical image sequences. The segmentation of the images is performed via a neural network, island removal and some manual adjustments with the ModaVox's toolboxes. An implementation of the TLM method allows for the simulation of the acoustic propagation of the input signal through the TLM mesh of the vocal tract model. ModaVox also generates tetrahedral and or surface triangle meshes.

A. S. Brandão
Mechanical Engineering Post-Graduation Program, Universidade Federal Fluminense, Rua Passo da Pátria 156, Bloco E, sala 216, CEP, São Domingos Niterói, RJ 24210-240, Brazil
e-mail: abrand@operamail.com

F. R. Leta
Mechanical Engineering Department, Mechanical Engineering Post-Graduation Program, Universidade Federal Fluminense, Rua Passo da Pátria, 156, bloco D, sala 302, São Domingos, Niterói, RJ 24210-240, Brazil
e-mail: fabiana@ic.uff.br

E. Cataldo (✉)
Applied Mathematics Department, Post-Graduation Program in Telecommunications Engineering, Universidade Federal Fluminense, Rua Mário Santos Braga, S/N Valonguinho, Niterói, RJ 24.020-140, Brazil
e-mail: ecataldo@im.uff.br

A. Öchsner et al. (eds.), *Design and Analysis of Materials and Engineering Structures*, Advanced Structured Materials 32, DOI: 10.1007/978-3-642-32295-2_12,

1 Introduction

Mesh extraction is an important step in any numerical scheme. Among the several existent 3D mesh extraction techniques, none is devoted to the extraction of meshes for the TLM method. Most of them are dedicated to extract tetrahedral or hexahedral meshes [1–3] for the finite element method (FEM) or surface meshes consisting of triangles [4] (or quadrilaterals) for the boundary element method (BEM).

Despite 3D TLM models can be seen in several works [5, 6], their geometry tends to be rather rectangular, because the problem of mesh generation for 3D TLM meshes has not been addressed yet. For example, the TLM model in [6] consists of a 3D vocal tract model represented by a sequence of ducts with rectangular cross-sectional areas, whose values were obtained by magnetic resonance images (MRI) measurements.

This chapter presents a simple mesh extraction technique, which uses the 3D grid structure formed by the voxels (volumetric picture elements) of a segmented object in a 3D image to build the edges and nodes of a 3D TLM mesh. The technique is implemented in the ModaVox (an open source software developed during this research) and will be explained and demonstrated in synthetic 3D images forming tube concatenation models of the vocal tract and also in a 3D MRI sequence of a real human vocal tract.

The only requirement regarding the presented mesh extraction technique is that to generate uniform 3D TLM meshes the 3D image must be isotropic (with regularly spaced voxels). In anisotropic 3D images, the edges of the TLM mesh will not be of uniform size.

This chapter is organized as follows: Sect. 2 describes the ModaVox application. Section 3 explains how the meshes were constructed. Section 5 shows some examples of model meshes and in Sect. 6 conclusions are outlined.

2 The Modavox Program

The ModaVox program was developed during this research. Its main objective is to study acoustic propagation of vowel sounds in 3D vocal tract models using the transmission line matrix (TLM) method. The name ModaVox stands for "Modelador da Voz" in Portuguese, or voice modeler in English. Currently, the ModaVox is able to: (i) select the volume of interest (VOI) (ii) segment the 2D and 3D images (iii) extract the surface (composed by triangles), tetrahedral and 3D grid TLM meshes (which is being discussed here) (iv) automatically detect the points pertaining to the frontiers of the extracted meshes (v) set boundary conditions to specific mesh points, including the definition of input points for the driven excitation signals, and (vi) apply the TLM method to the meshes producing output (*.csv) files. Its implementation includes four open source components (Qt

[7], VTK [8], ITK [9] and TetGen [10]. A self-organizing map neural network and the TLM method were implemented independently.

The segmentation of the images is performed via a neural network, island removal and, in some cases, manual segmentation adjustments with the ModaVox's toolboxes. At the present moment, an implementation of the TLM, with the lossless shunt node, allows for the simulation of the acoustic propagation of input signals through the TLM mesh models.

A more complete introduction about the ModaVox can be seen in [11].

3 Mesh Extraction

The TLM meshes are constructed over the voxels (volumetric picture elements) that have the object label (or gray level) from the segmented 3D images. Consequently, the extracted TLM mesh will fit the segmented object as closely as possible, depending only on the size of the voxels. As this mesh extraction methodology is dependent upon the resolution of the 3D image, the TLM numerical dispersion must be considered in order to define this resolution.

In the TLM method, at each time step, every node receives incident voltage (pressure) pulses and sends scattered pulses, generated using the scattering matrix, Fig. 1.

The main limitation of TLM is the dispersion error, because the velocity of a propagating wave is dependant upon both its frequency and direction of travel, leading to wave propagation errors and a mistuning of the expected resonance frequencies. The degree of dispersion error is highly dependant upon mesh topology and has been investigated in many works [12, 13]. In the work of Murphy et al. [14], it is mentioned that the minimization of dispersion can be achieved through the use of interpolated or dodecahedral mesh topologies. However, 3D grid meshes are easier to construct and less computationally expensive. Other approaches tried to solve the numerical dispersion problem by doing pre and post processing of results from the mesh structures for posterior application of frequency warping techniques in order to correct mistuned resonance frequencies [15, 16].

The main numerical dispersion minimization technique in TLM is to set the discrete spatial step (ΔL) much smaller than the wavelength (λ), according to Eq. (1).

$$\Delta L/\lambda \leq 0.1 \tag{1}$$

which ensures that the difference of the propagation velocity between waves of different wavelengths will be small (Fig. 2).

According to the plot in Fig. 2, if $\Delta L = 1$ mm or less, the frequency-dependent numeric dispersion is sufficiently reduced. Hence, according to Eq. (1), the highest frequency that a signal component can have in order to travel through the TLM mesh is 34,310 Hz for the tube meshes.

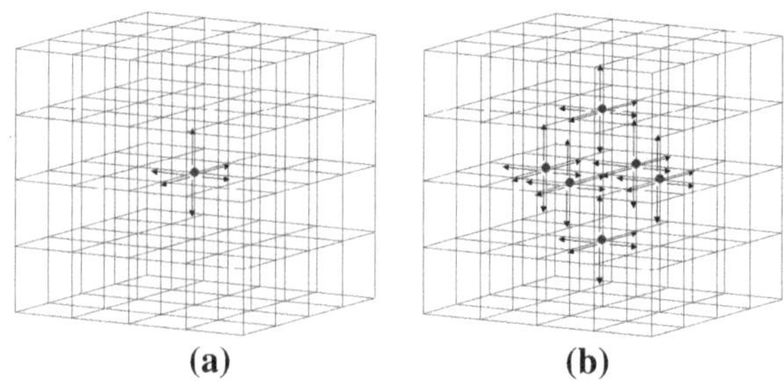

Fig. 1 The 3D TLM. **a** Pulses propagation at instant t **b** Pulses propagation at instant t + Δt

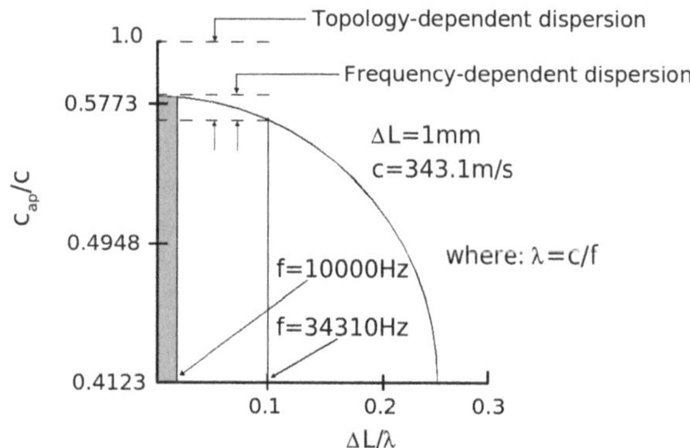

Fig. 2 Numeric dispersion in 3D TLM (adapted from [22])

Once ΔL is defined, the size of the voxels is defined by (ΔL, ΔL, ΔL). The 3D image must be isotropic, in order to obtain a uniform mesh. However, the process can be applied to any 3D image. The only difference is that in anisotropic images, as some 3D MRI sequences, the edges of the TLM mesh will not be of uniform size.

The time discretization, Δt, can be defined by Eq. (2):

$$\Delta t = \Delta L \big/ (c\sqrt{D}), \tag{2}$$

where c is the air propagation sound velocity and $\sqrt{D}$ is the distance between source and target nodes in a 3D grid uniform TLM mesh. The factor $\sqrt{D}$, which will not be discussed here, is necessary to compensate for the topology-dependent numeric dispersion (see Fig. 2) considering 3D grid uniform mesh topology.

3.1 Tube Model Meshes

The process of obtaining the tube meshes is performed in four steps: (i) The slice sequence for the tube mesh generation is constructed from a single drawing (here, the Kolourpaint [17] image editor was used to draw the first slice of the tube);

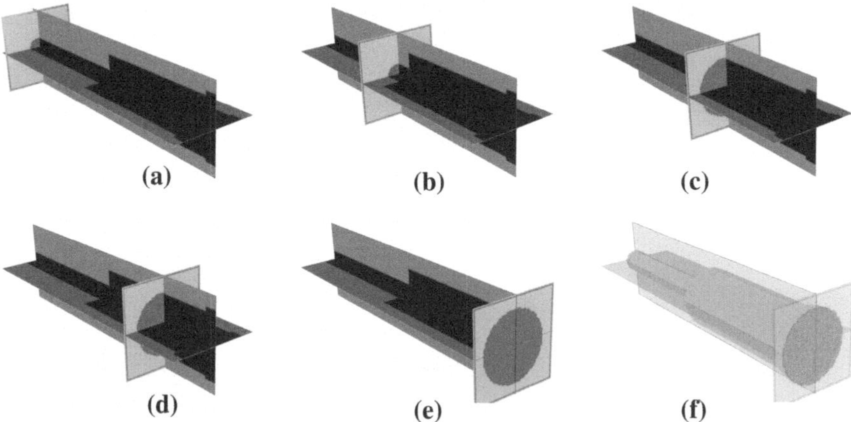

Fig. 3 3D TLM mesh extraction for the two-tube concatenation model. **a–e** Segmentation of the volume image **f** Construction of the TLM mesh over the segmented voxels

(ii) The single image is saved in the DICOM format with the Gimp [18] image editor; (iii) Several copies of the DICOM slice are created and then the ModaVox software is used to open and append this slice sequence into a volume image with the desired tube length; and (iv) The ModaVox is used again to segment the several slices of the created volume image, each one with a single disk object, appending them in a segmented 3D volume image from which the TLM mesh is extracted.

To construct the two-tube model, shown in Fig. 3, mesh with sections of different diameters, two slice sequences should be created, as described above. For all tube models, according to Eq. (1) the space discretization ΔL is equal to 1 mm.

Due to its simplicity and absence of noise, the slices of the tube sequences are segmented automatically by the ModaVox. However, as shown in the following subsection, the slices for the 3D MRI sequence for the human vocal tract need some manual segmentation adjustments.

3.2 Vocal Tract Mesh

The process of generating the vocal tract mesh is performed in four steps: (i) The MRI 3D sagittal sequence is extracted from a subject for the /a/ vowel shape, with an acquisition time of 18 min 39 s. The subject emits the /a/ vowel periodically, helping himself to maintain the vocal tract in a fixed position during the extraction of the sequence, which was originally composed of 186 slices of 512 × 512 pixels. The sequence was acquired using a General Electric machine, model GE Medical Systems Signa HDxt, with a 1.5 Tesla magnetic field strength, using the following

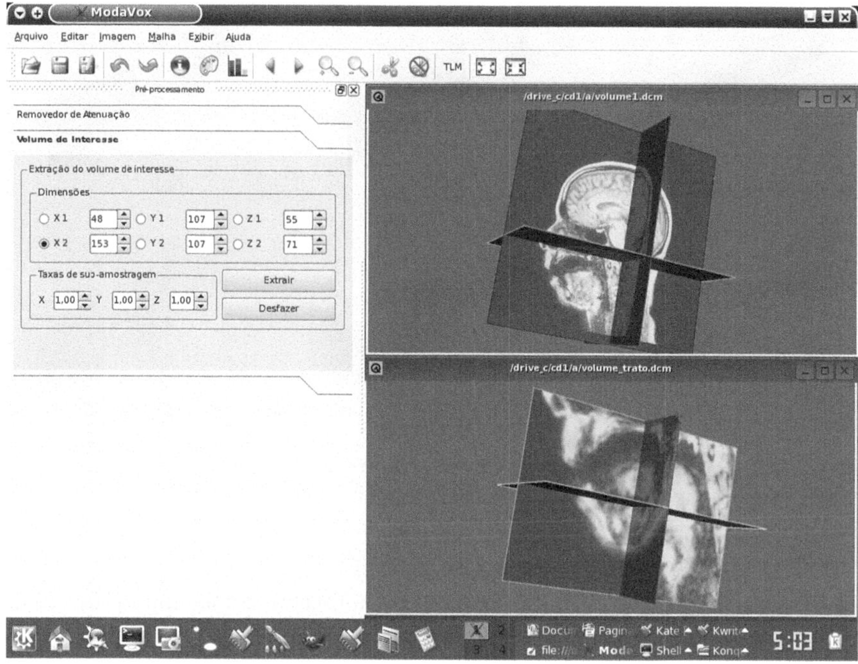

Fig. 4 Selection of the volume of interest (VOI) with the ModaVox program

parameters: T1-weighted MR image, gradient recalled (GR) scanning sequence, 8 channel head coil (8HRBRAIN), echo time of 4.744 ms, repetition time of 10.584 ms, field of view (FOV) 48 $\times$ 48 cm, pixel spacing (0.9375, 0.9375) mm and 1 mm slice thickness with no gap, hence, an anisotropic image; (ii) The volume of interest (VOI) dimensions are passed, through the ModaVox's interface, to an object of the class vtkExtractVOI from the VTK [8] library, which extracts the VOI (see Fig. 4), reducing the original volume image, which is oversampled to remove the anisotropy. Hence, the final voxel dimensions were (0.968498, 0.968498, 0.968498) mm; (iii) On this new isotropic 3D image, the segmentation is performed via a neural network, islands removal and some manual adjustments (all of these operations were performed using the ModaVox program). The teeth part was segmented manually. (iv) Finally, the uniform TLM mesh is extracted from the segmented 3D image.

3.3 The Mesh Extraction Algorithm Object Pipeline

The code for the mesh extraction algorithm is based on the object pipeline shown in Fig. 5.

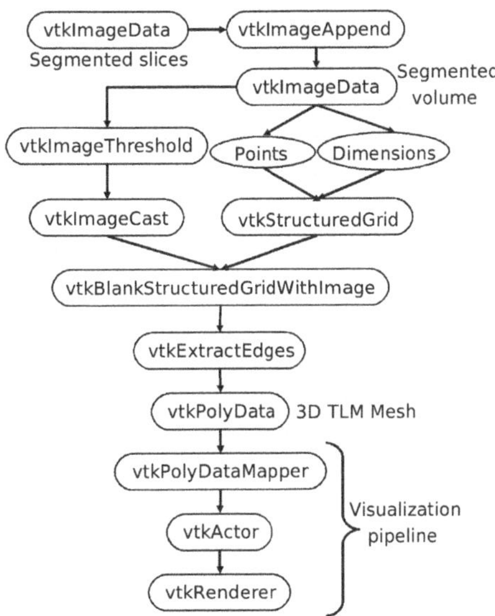

Fig. 5 Mesh extraction algorithm object pipeline. The classes that compose the object pipeline are explained in the VTK library documentation [8]

4 Mesh Attributes

At each mesh node, the pressure value represents the solution of the acoustic wave equation, at the current iteration and the boundary values define the boundary conditions which will be applied, setting specific values for the reflection (ρ) and transmission (τ) coefficients of the transmission lines. The boundary values and their corresponding meanings are: (i) Value 0—interior nodes ($\rho = -2/3$ and $\tau = 1/3$); (ii) Value 1—reflective condition ($\rho = 1$ and $\tau = 0$); (iii) Value 2—input nodes (also interior nodes); (iv) Value 3—free space condition ($\rho = -1$ and $\tau = 0$).

5 Mesh Examples

5.1 Single Tube TLM Meshes

In voice and speech research, tube models are frequently used to approximate the shape of the vocal tract [19, 20]. The single tube models meshes are the simplest example, as shown in Fig. 6.

5.2 Two-Tube Model for the Vocal Tract Simulating an Open /a/ Vowel

Let L_1, L_2, A_1 and A_2 be, respectively, the lengths and cross-sectional areas of the two tube sections. According to Stevens [21], to construct the two-tube model mesh for the /a/ vowel, the values must be $L_1 = 90$ mm, $L_2 = 80$ mm, $A_1 = 100$ mm^2,

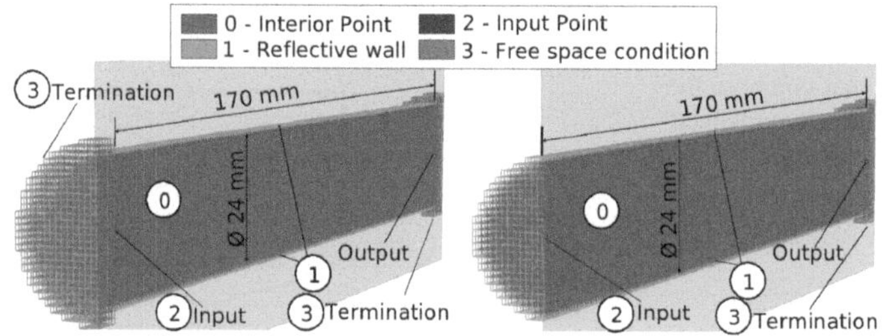

Fig. 6 Meshes for tubes with 24 mm in diameter and 170 mm in length **a** Open tube **b** Closed tube

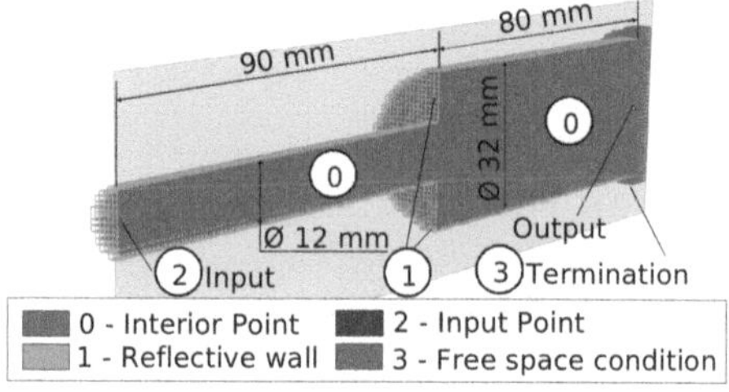

Fig. 7 Two-tube model for the /a/ Vowel

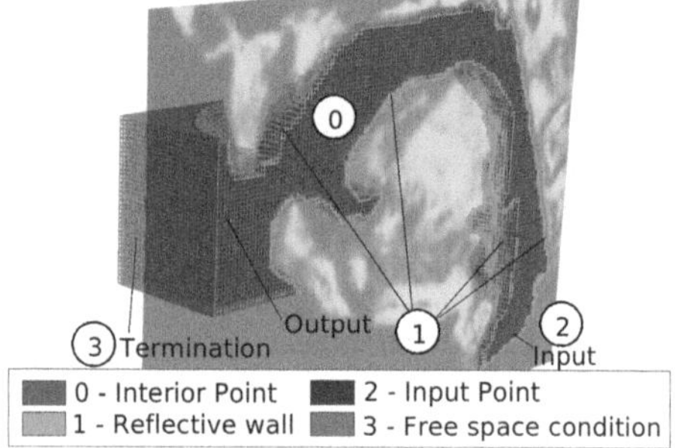

Fig. 8 Vocal tract mesh corresponding to an /a/ vowel shape. The cut plane shows a sagittal slice from the MRI sequence

$A_2 = 700\ \text{mm}^2$. However, due to the approximation of the smaller diameter $(d_1 = 2\sqrt{A_1/\pi})$ to $d_l = 12$ mm, the diameter of tube 2 (d_2) was approximated to 32 mm (see Fig. 7), so that the ratio between the original values of d_2 and d_l, and, consequently, between the cross-sectional areas, remains closely the same, i.e.,

$$\frac{d_2}{d_1} = \frac{29.85410660}{11.28379167} = 2.6457 \cong 2.6666 = \frac{32}{12} \tag{3}$$

The two-tube model for the /a/ vowel is shown in Fig. 7.

An animation sequence for the extraction of the mesh in Fig. 7 is shown in Fig. 3.

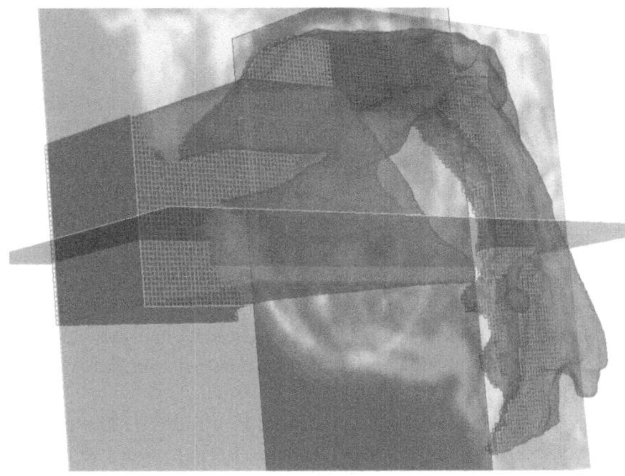

Fig. 9 Comparison between the 3D TLM and a triangle surface (in translucent surface visualization) meshes extracted for the same model segmented object

5.3 *The 3D TLM Human Vocal Tract Mesh*

As an application of the methodology proposed, a human vocal tract TLM mesh whose shape corresponds to the /a/ vowel was extracted. Magnetic resonance images were used for the 3D reconstruction of the vocal tract, as explained in Sect. 3.2. For the vocal tract mesh, shown in Fig. 8, $\Delta L = 0.968498$ mm due to anisotropy removal.

Figure 9 shows a comparison between the surface mesh, extracted with the Marching Cubes algorithm, and the TLM mesh for the same segmented volume of the vocal tract. The two meshes are superposed over the same segmented object.

6 Conclusion

A mesh extraction technique for high quality 3D TLM models was developed. It can generate 3D TLM models from any segmented object in a volume image, because it fits the object as closely as possible, depending only on the size of the voxels. This was demonstrated with the tubes and with the human vocal tract model.

In addition to build extract the meshes, the developed open source application (ModaVox) can apply boundary conditions and execute the TLM method.

Acknowledgments The authors thank Dr. Alair Augusto S.M.D. dos Santos from Hospital das Clínicas de Niterói for the valuable Radiology suggestions and for making the MR equipment of the ProEcho clinic available to this research. This work was supported by FAPERJ (Fundação Carlos Chagas Filho de Amparo a Pesquisa no Estado Rio de Janeiro), by CAPES (Coordenação de Aperfeiçoamento de Pessoal de Nível Superior) and by CNPq (Conselho Nacional de Desenvolvimento Científico e Tecnológico).

References

1. Zhang, Y., Bajaj, C.: Adaptive and quality quadrilateral/hexahedral meshing from volumetric data. Comput. Methods Appl. Mech. Eng. **195**(9–12), 942–960 (2006)
2. Zhang, Y., Bajaj, C., Sohn, B.S.: 3d finite element meshing from imaging data. Comput. Methods Appl. Mech. Eng. **194**(48–49), 5083–5106 (2005)
3. Zhang, Y., Hughes, T.J., Bajaj, C.L.: An automatic 3d mesh generation method for domains with multiple materials. Comput. Methods Appl. Mech. Eng. **199**(5–8), 405–415 (2010)
4. Briaire, J.J., Frijns, J.H.M.: 3d mesh generation to solve the electrical volume conduction problem in the implanted inner ear. Simulat. Pract. Theory **8**(1–2), 57–73 (2000)
5. Amri, A., Saidane, A., Pulko, S.: Thermal analysis of a three-dimensional breast model with embedded tumour using the transmission line matrix (TLM) method. Comput. Biol. Med. **41**(2), 76–86 (2011)
6. El-Masri, S., Pelorson, X., Saguet, P., Badin, P.: Development of the transmission line matrix method in acoustics. Application to higher modes in the vocal tract and other complex ducts. Int. J. Numer. Model. **11**(3), 133–151 (1998)
7. Blanchette, J., Summerfield, M.: C++ GUI Programming with Qt 4. Prentice-Hall, Englewood Cliffs (2006)
8. Schroeder, W., Martin, K., Lorensen, B.: The visualization toolkit: an object-oriented approach to 3-D graphics, 3rd edn, Kitware Inc., Clifton park (2002)
9. Ibanez, L., Schroeder, W.: The ITK Software Guide 2.4, 2nd edn. Kitware Inc., clifton park (2005)
10. Si, H.: Tetgen: a quality tetrahedral mesh generator. http://tetgen.berlios.de(2002). Accessed 20 Jan 2008
11. Brandão, A.S.: The Modavox software (Announced at Professor Edson Cataldo's homepage). http://www.professores.uff.br/ecataldo/ index.php?option=com_content&view=article&id=33&Itemid=49(2010). Accessed 03 Sept 2011
12. Campos, G.R., Howard, D.M.: On the computational efficiency of different waveguide mesh topologies for room acoustic simulation. IEEE Trans. Speech Audio Process. **13**(5), 1063–1072 (2005)
13. Fontana, F., Rocchesso, D.: Signal-theoretic characterization of waveguide mesh geometries for models of two-dimensional wave propagation in elastic media. IEEE Trans. Speech Audio Process. **9**(2), 152–161 (2001)
14. Murphy, D., Kelloniemi, A., Mullen, J., Shelley, S.: Acoustic modeling using the digital waveguide mesh. IEEE Signal Process. Mag. **24**(2), 55–66 (2007)
15. Fontana, F.: Computation of linear filter networks containing delay-free loops, with an application to the waveguide mesh. IEEE Trans. Speech Audio Process. **13**(5), 774–782 (2003)
16. Savioja, L., Välimäki, V.: Interpolated rectangular 3-d digital waveguide mesh algorithms with frequency warping. IEEE Trans. Speech Audio Process. **11**(6), 783–789 (2003)
17. Dang, C.: Kolourpaint is a free, easy-to-use paint program for KDE. http:// kolourpaint.sourceforge.net/ (2003). Accessed 16 Oct 2006
18. Mattis, P., Kimball, S.: Gnu image manipulation program. http://www.gimp.org/ (1995). Accessed 13 May 2007
19. Kagawa, Y., Tsuchiya, T., Fujii, B., Fujioka, K.: Discrete Huygens' model approach to sound wave propagation. J. Sound Vib. **218**(3), 419–444 (1998)
20. Rabiner, L.R., Schafer, R.W.: Digital Processing of Speech Signals. Prentice-Hall, Englewood Cliffs, Chapter 3 (1978)
21. Stevens, K.: Acoustic Phonetics. MIT Press, Cambridge (1998)
22. Cogan, D., O'Connor, W., Pulko, S.: Transmission Line Matrix in Computational Mechanics. CRC Press, Taylor & Francis Group, Boca Raton, Florida, pp. 102–104 (2006)

Index

A. Öchsner et al. (eds.), *Design and Analysis of Materials and Engineering Structures*,
Advanced Structured Materials 32, DOI: 10.1007/978-3-642-32295-2,

The manufacturer's authorised representative in the EU is Springer Nature Customer Service Centre GmbH, Europaplatz 3, 69115 Heidelberg, Germany. If you have any concerns regarding our products, please contact ProductSafety@springernature.com

Printed and bound by CPI Group (UK) Ltd, Croydon, CR0 4YY
15/07/2026
02167647-0001